72nd Annual Meeting of the Labor and Employment Relations Association and LERA@ASSA Meeting 2020

Labor and Employment Relations Association Series

San Diego, California
3-5 January 2020

Online
13-16 June 2020

Editors:

Ariel Avgar
J. Ryan Lamare

ISBN: 978-1-7138-2424-4

Printed from e-media with permission by:

Curran Associates, Inc.
57 Morehouse Lane
Red Hook, NY 12571

Some format issues inherent in the e-media version may also appear in this print version.

Printed with permission by Curran Associates, Inc. (2021)

For permission requests, please contact Labor and Employment Relations Association (LERA) at the address below.

Labor and Employment Relations Association (LERA)
University of Illinois at Urbana-Champaign
121 Labor and Employment Relations Bldg.,
504 E. Armory Ave.
Champaign IL 61820
USA

Phone: (217) 333-0072
Fax: (217) 265-5130

LERAoffice@illinois.edu

Additional copies of this publication are available from:

Curran Associates, Inc.
57 Morehouse Lane
Red Hook, NY 12571 USA
Phone: 845-758-0400
Fax: 845-758-2633
Email: curran@proceedings.com
Web: www.proceedings.com

LABOR AND EMPLOYMENT RELATIONS ASSOCIATION SERIES

Proceedings of the LERA 2020 Meetings

LERA@ASSA Meeting
January 3-5, 2020
San Diego, California

LERA 72nd Annual Meeting
June 13-16, 2020
Virtual

OFFICERS OF LERA FOR 2019–2020

PRESIDENT

Dennis Dabney, Kaiser Permanente

PRESIDENT ELECT

Adrienne Eaton, Rutgers University

PAST PRESIDENT

Kris Rondeau, AFSCME

SECRETARY–TREASURER

J. Ryan Lamare, University of Illinois at Urbana-Champaign (January 2019–June 2020)
Andrew Weaver, University of Illinois at Urbana-Champaign (June 2020–)

EDITOR-IN-CHIEF

Ariel C. Avgar, Cornell University (overlapping 1 year)
J. Ryan Lamare, University of Illinois at Urbana-Champaign (overlapping 1 year)

CHAPTER ADVISORY CHAIR

William Canak, Middle Tennessee State University

EXECUTIVE DIRECTOR

Emily Smith, University of Illinois at Urbana-Champaign

REGIONAL VICE PRESIDENTS

Jim Pruitt, Kaiser Permanente (West Region)
Michele Hoyman, University of North Carolina at Chapel Hill (East Region)
Robert Chiaravalli, Strategic Labor and HR LLC (Mid Region)

EXECUTIVE BOARD MEMBERS

Sylvia Allegretto, Center on Wage & Employment Dynamics
Daniel Altchek, Miles & Stockbridge, P.C.
John Amman, IATSE Local 600
Ezio Borchini, George Washington University
Heather Boushey, Washington Center for Equitable Growth
Paul Clark, Penn State University
Bill Dirksen, Ford Motor Company
Virginia Doellgast, Cornell University
Cyndi Furseth, Portland General Electric (ret.)
Kati Griffith, Cornell University
Erin Johansson, Jobs With Justice
Brad Markell, Industrial Union Council (AFL-CIO)
Sheila Mayberry, Arbitrator/Mediator
Jake Rosenfeld, Washington State University, St. Louis
Maite Tapia, Michigan State University
David Weil, Brandeis University

LEGAL COUNSEL

Steven B. Rynecki, von Briesen, Purtell & Roper, S.C.

PAST PRESIDENTS OF LERA

1948–Edwin E. Witte
1949–Sumner H. Slichter
1950–George W. Taylor
1951–William Leiserson
1952–J. Douglas Brown
1953–Ewan Clague
1954–Clark Kerr
1955–Lloyd G. Reynolds
1956–Richard A. Lester
1957–Dale Yoder
1958–E. Wight Bakke
1959–William Haber
1960–John T. Dunlop
1961–Philip Taft
1962–Charles A. Myers
1963–William F. Whyte
1964–Solomon Barkin
1965–Edwin Young
1966–Arthur M. Ross
1967–Neil Chamberlain
1968–George P. Shultz
1969–Frederick Harbison
1970–Douglass V. Brown
1971–George Hildebrand
1972–Benjamin Aaron
1973–Douglas H. Soutar
1974–Nathaniel Goldfinger
1975–Gerald G. Somers
1976–Irving Bernstein
1977–F. Ray Marshall
1978–Charles C. Killingsworth
1979–Jerome M. Rosow
1980–Jack Barbash
1981–Rudolph A. Oswald
1982–Milton Derber
1983–Jack Stieber
1984–Wayne L. Horvitz
1985–Everett Kassalow
1986–Lloyd Ulman
1987–Michael H. Moskow
1988–Phyllis A. Wallace
1989–Joyce D. Miller
1990–Robert B. McKersie
1991–James L. Stern
1992–Ernest J. Savoie
1993–George Strauss
1994–Lynn R. Williams
1995–Walter Gershenfeld
1996–Hoyt N. Wheeler
1997–Francine D. Blau
1998–F. Donal O'Brien
1999–Thomas A. Kochan
2000–Sheldon Friedman
2001–Magdalena Jacobsen
2002–John F. Burton, Jr.
2003–Paula B. Voos
2004–Marlene Heyser
2005–Stephen R. Sleigh
2006–David B. Lipsky
2007–Eileen B. Hoffman
2008–Anthony Oliver, Jr.
2009–Joel Cutcher-Gershenfeld
2010–Eileen Appelbaum
2011–Gordon Pavy
2012/13–David Lewin
2013/14–Lisa Lynch
2014/15–Martin Mulloy
2015/16–Bonnie Castrey
2016/17–Janice Bellace
2017/18–Harry Katz
2018/19–Kris Rondeau

LABOR AND EMPLOYMENT RELATIONS ASSOCIATION SERIES

Proceedings of the LERA 2020 Meetings

LERA@ASSA Meeting
January 3–5, 2020, San Diego, CA
(in conjunction with ASSA/AEA)

and

LERA 72nd Annual Meeting,
June 13–16, 2020, Virtual

Ariel Avgar, Co-Editor-in-Chief
J. Ryan Lamare, Co-Editor-in-Chief

LABOR AND EMPLOYMENT RELATIONS ASSOCIATION SERIES
(Formerly the Industrial Relations Research Association)

Proceedings of the Annual Meeting
Annual Research Volume
Perspectives on Work Magazine

Inquiries and other communications regarding membership, meetings, publications, and general affairs of the Association, as well as notice of address changes, should be addressed to the LERA office.

LABOR AND EMPLOYMENT RELATIONS ASSOCIATION
University of Illinois at Urbana-Champaign School of Labor and Employment Relations
121 Labor and Employment Relations Building, MC-504
504 East Armory Avenue, Champaign, IL 61820
Telephone: 217/333-0072 Fax: 217/265-5130 E-mail: LERAoffice@illinois.edu
Website: http://lera.memberclicks.net

CONTENTS

LERA 72ND ANNUAL MEETING, JUNE 13–16, 2020, VIRTUAL

I. Presidential Address

The Advantages of Partnership in Labor Relations

DENNIS DABNEY
Kaiser Permanente

To everyone who helped us carry out LERA's 72nd annual meeting in a virtual platform—thank you for coming together with us to attend to the important business at hand.

It is nothing short of phenomenal that we were able to pull it off!

The experience leaves me more eager than ever to talk about partnership, as practiced within the Labor Management Partnership at Kaiser Permanente. Our organization views partnership as more than a labor relations strategy; it is a business strategy that brings together multiple perspectives—from managers to front-line employees—to inform important decisions and drive innovation.

One of the biggest advantages to partnership, in my mind, is the way it can foster strong relationships between labor and management. These relationships can help move a large organization through unexpected challenges.

This certainly has been true with the COVID-19 pandemic, where Kaiser Permanente's engagement of our many employee unions was vital in developing a comprehensive plan for the expected surge in patients caused by the spread of the novel coronavirus.

The advantages of partnership resonate with important lessons about labor and management relationships I learned early in my career.

Management chose me for this work by taking out the magic wand, dubbing me "labor relations," and sitting me in the LR department of a large automobile parts manufacturer plant in Detroit.

I had zero training or understanding of this field, but the school of hard knocks was patiently waiting to teach me.

One day I received a telephone call in my office asking me to hurry out to the press shop.

That meant stepping into a large shop dominated by the loud thunder of automotive presses. These presses are three stories tall, with eight presses in each roll. The machinery made things like an extended 13-foot Chrysler van roof.

The first thing I noticed as I walked into the shop was that there was no noise. It was eerily quiet.

On the shop floor, I saw a manager and a union representative standing at a lead press with their arms folded, facing off. All the employees had stopped working and were watching the two men.

Not knowing what to do in this situation, my LR instincts kicked in. I invited the two into an office to talk. They both readily agreed and, as we walked off, I could hear the presses beginning to run in the background.

That was a good sign, but I still had no idea how I was going to resolve the dispute. Luckily, when we got to the office, I didn't need to say a word.

That's because these two guys began apologizing to each other for being put in this situation.

On the open shop floor, in front of all those workers, neither man felt he could back down or defuse the situation. They didn't want to appear weak or not in control.

EARLY LESSONS

I tell this story because it was my first lesson in understanding that management and labor both want to work together to solve problems, produce quality products, and provide a voice and security for workers.

I later built on these on-the-job lessons by taking labor relations classes at Cornell, Michigan, Harvard, and other prominent schools.

I've also had the privilege of working with some of the finest university professors in the country, such as Joel Cutcher-Gershenfeld at Brandeis, Thomas Kochan at MIT, and Harry Katz at Cornell.

I've had many opportunities to learn and to grow in this field. But firsthand experience has always been the best teacher of them all.

The second story I want to tell came later, just when I was feeling ready to make my mark in this field. I was part of a team that hosted representatives from the United Auto Workers for an important bargaining session. I brought everything I had to this meeting—except my common sense and humility.

I remember feeling pretty good when the union listened to our first set of demands—yes, they were demands—and asked legitimate, probing questions.

Later, a great UAW leader named Herb Wilbert pulled me aside and said, "Dennis, I noticed you guys had a full spread buffet in your caucus room—coffee, donuts, fruit, the works."

Mr. Wilbert noted the same was not true for the union caucus room, which was supplied only with pitchers of water. He made the point that as hosts we clearly were not treating them as we treated ourselves.

The lesson here was not about the coffee or donuts.

It was about relationships and bargaining, and the importance of trust … integrity … and being a person of your word—not to mention treating others as you want to be treated.

COMMON INTERESTS

I have one more story.

This is from later in my career, after I moved on to the utility industry, working with leaders such as Mike Langford from the utility workers union and Jim Hunter from the IBEW. It was at this time that I learned more about interest-based bargaining and a facilitated process.

One day, we're at the bargaining table negotiating over—what else?—benefits, and we had an interest on the table that employees participate in the cost of health benefits. A union representative asked the company, in a not-so-pleasant way, "How much money do we want employees to give from their paychecks?"

We agreed that was a good question. Many days later, we settled for much less than the 5 percent originally asked for. For me, that decision brought an end to the days of arguing over the expenses of healthcare.

The lesson I learned here was that an interest-based conversation can deliver better outcomes for all parties involved.

These experiences—and the lessons they carried—continue to inform my work in healthcare and, more specifically, with labor–management partnerships.

They also illustrate why I think partnership is a better approach to labor relations.

In making this case, I need to be clear: Partnership brings its challenges. It's not Utopia.

But I think the challenges that partnership present are worth the trouble because partnership expands the opportunities for developing strong relationships between labor and management.

A cornerstone of a great partnership is teams of employees working together to achieve better performance, quality, and customer satisfaction—which results in employees getting a voice in the work environment and job security.

At Kaiser Permanente, much of this work occurs within the more than 3,600 unit-based teams—UBTs—working across the organization. These teams drive efforts to continually improve the care we provide to our 12.3 million members and to also improve the workplace experience of our tens of thousands of employees.

The work of our UBTs affects virtually every aspect of our business.

We have teams engaged in continuous improvement and innovation—not only in clinical settings but in our emergency departments, maternity, pharmacy, pediatrics, and oncology units. And we have these teams in support settings, from food and nutrition services to our housekeeping and janitorial staff to medical records units.

GETTING RESULTS

And they clearly are making a difference. On a regular basis, we evaluate how well our teams are working as teams. That information, combined with quarterly surveys of Kaiser Permanente employees, provides strong evidence of the positive impact teams have; for example,

- The surveys show that employees closely involved in UBT work have significantly better scores in response to questions about their health and overall well-being.
- These employees are also much more likely to feel they have voice in making important decisions about their work and can speak up safely on the job. This is an extremely important point because it ties directly to patient safety.
- Departments with high-performing UBTs also have significantly improved results: They have fewer workplace injuries. They have fewer lost workdays. They see improved patient satisfaction.

The success of our teams reflects the truth about labor–management relationships that I learned early in my career.

Many experiences I've had since convince me that the partnership approach we pursue is a better way to go.

Not easy. But better.

Over the years, I've worked in labor relations in the traditional model, the more cooperative model, and the model that informs our Labor Management Partnership—the interest-based model.

I prefer interest-based partnership because I think it helps achieve superior outcomes for the employer, the employees, and the unions.

Partnership brings together diverse points of view, which leads to better decisions.

This approach to doing business requires daily collaboration between labor and management. Sometimes the process is frustrating. But the frequent collaboration builds strong, ongoing relationships.

The collaboration fosters trust.

And it supports a collective belief that it's possible for labor and management to both win.

It doesn't guarantee labor–management harmony—the differences between labor and management are inevitable, especially during bargaining.

But years in this business have convinced me that it is vitally important to let the shared interests—not differences—define the relationship.

BOTH SIDES CAN WIN

My experience in partnership has shown me that it's possible to develop labor–management relationships that are about much more than who gets a larger slice of the pie. Partnership has proven to be a winning strategy for all parties:

- We have seen continuous improvement in service, quality, and performance.
- At a time when union membership has declined in the United States, our partnership unions have seen their membership almost double in the past two decades.
- Our workers consistently receive industry-leading wages and benefits, commitments to job security and a powerful voice in key decisions.

I will close with a quote from our former CEO, Bernard Tyson, who passed away in November of last year and whose support for our partnership was unrelenting.

"Our Labor Management Partnership has helped make Kaiser Permanente a better place to get care, and a better place to work. It has made Kaiser Permanente and its workforce stronger, set us apart, and given KP a unique competitive advantage."

Maybe it can do the same for your organization.

II. A Second Look at Sectoral Bargaining

Rethinking Multi-Employer Benefit Structures for Applicability to Today's Labor Markets

HEATHER L. GROB
Saint Martin's University

Labor organizations whose members are subject to greater numbers of work breakages or variable hours traditionally have relied on multi-employer benefit structures to provide greater long-term security for employees. In the United States, these structures are most often jointly managed between labor unions and management associations (Taft–Hartley funds), but other types of multiple employer plans have also grown in popularity. Institutions with sectoral and multi-employer approaches have been able to collectively bargain at higher and sustained levels than have institutions that use enterprise-level approaches, when viewed through an international lens (OECD 2017). But multi-employer benefit structures have also been under financial siege in the past several decades owing to a decline in unionization and pay rates. Labor in multi-employer contexts faces a crossroads: argue for state intervention for continuity of employee benefits or work toward privately managed labor–management systems to secure jobs, pay, and benefits for their members.

For the sake of clarity, let me define that multi-employer benefits or structures (*multi's*) would mean those generally allowable under the National Labor Relations Act of 1947 (Taft–Hartley Act), which expressly allowed funds such as pensions and health and welfare funds to be jointly managed and enforced as part of a collective bargaining agreement. The joint labor–management entity often contracts with a third-party benefit provider to offer benefits or training to workers so that they may change employers or jobs in particular sectors without losing benefits and while preserving a pool of workers with necessary skills. I will refer to *multiple* employer plans or strategies (MEPS) as those attempts by either unions or businesses to develop plans that allow for portable benefits or skill development. Both Multi's and MEPS are often administered by third parties and targeted to particular sectors, but MEPS do not require an agreement through collective bargaining to legitimize the arrangement.

In this paper, I discuss multi-employer jointly managed pension, health, and welfare funds as well as apprenticeship, training programs, hiring halls, and dispute resolution systems as primary areas for labor policy focus with a view toward structures that could be developed to support so-called gig work. While I cannot possibly go into all of the challenges these multi- or multiple employer arrangements present, the intent is to provide a broad sweep of issues for consideration with a view toward legal changes that could help to improve the situation of those who find themselves working but unable to obtain benefits for healthcare, retirement, continuing training, and education.

Who Might Benefit from Expansion of a Multi-Employer System

The technological, political, and economic forces that drive increased gig work are also driving the need to rethink employer benefit systems. For instance, many building trade occupations are contingent by the inherent nature of the work itself: work is often temporary because the workplace is usually equivalent to the product, and weather, finance, coordination, and logistics of complexities involved in construction can create breakages in continuous workflow, but the work is most done domestically. In some sectors, skilled technicians are in high demand for short periods of time at different workplaces—for instance, healthcare or emergency workers who may be required to respond to an epidemic or war in a location separate from their home or employer's location. The same could also be said for several performing arts organizations, such as theater, dance, and live music, where the quality of the product matters immensely. Delivery of people, goods, and materials to decentralized locations and geographically or seasonally varied work such as food production may also translate to varied workflow and demand for production or services—hence, volatile labor demand and therefore insecurity of employment-based benefits. Deregulation of trucking has certainly contributed to this problem, so the problems of contingency are not necessarily "natural" outcomes but a direct result of economic policies.

More recent drives toward "fissuring" have reached industries that heretofore could not fall in the above categories. For instance, cleaners and maintainers of buildings, along with food services and human resource departments that have fairly regular schedules have been contracted, subcontracted, and even sub-subcontracted to the point where it is difficult to know who the employer of responsibility is (Weil 2014). Prefabrication, 3D printing, robotics, and modular housing has also led to restructuring in parts of the construction sector. Certainly, the rise of digital production is changing the nature of work, so the intent here is to uncover parameters that may shape the way employee benefits potentially could be structured with changes in legal and institutional rules.

People lacking long-term access to single employer-oriented benefits might gain advantage from an expanded view of multi-employment benefits. Part-time or multiple job holding is more likely to occur in food preparation, education and training, art, design, entertainment and sports, and healthcare according to the Bureau of Labor Statistics (Berkhusen 2019). Interestingly, according to that study, construction does not appear to have a higher likelihood of multiple job holding than single job holding. However, over 3.4 million construction workers are also likely incorrectly classified as independent contractors, so the numbers of employees may be undercounted [U.S. General Accountability Office (GAO) 2009]. The GAO found that by a narrow definition of "core contingent workers," two thirds are less likely to have work-provided retirement and are less likely to be provided health insurance. Contingent workers are also more likely to live in poverty and rely on public assistance than are standard workers (GAO 2015: 29–31). The Bureau of Labor Statistics estimates that the number of low-wage jobs will grow significantly (Lerman 2019). Further, automation brings uncertainty to the job picture, making it harder to invest in occupational training for the longer term as whole sectors become replaced. Manyika et al. (2017) estimate that as much as one third of the workforce will need to change occupations by 2030.

The problem of developing and supporting a skilled workforce is a well-known public goods problem, but the solution may not be simply pay or government regulation. Just paying people more will attract individuals to jobs, certainly, but the organization required to train a workforce to acquire new skills while smaller employers vie for competitive hold is a problem of industrial organization and planning. Additionally, technology is changing rapidly, and regulations are slow to adapt. Fortunately, there is a general model for handling these problems. This model is not unique to building trades—many service sector jobs in the arts, healthcare, hotels, and restaurants also face similar issues.

Short-time work is held for a variety of reasons. Some gig workers want to make a little extra money in retirement, control their own schedules, or balance work with needs to care for school-age children or aging parents, or face health limitations. We can't assume, though, that these people do not need employer- or government-based benefits. Many could, and should, receive some sort of credit in the form of benefits, but their employers do not offer them for the perceived cost savings, lack of affordable solutions, or lack of access to an employer benefits model that would allow for short-term contributions. The principal problems of workers who may find themselves in a constant search for gigs or task work in the United States are that those jobs do not fit into a single-job and employment-based benefit system, which leads to further discriminant outcomes along the lines of gender, race, and ethnicity.

Retirement Plans

The retirement savings crisis has been well-documented. The foundation of retirement security in the United States is Social Security, but the private pension system, along with savings, is also very important for security in retirement. Over the past few decades, the American private pension system has shifted to account-based plans and has persistently excluded many workers. Account-based, defined-contribution plans are now the dominant type of plan over defined benefit plans. Nearly half of all Americans age 55 and older have no retirement savings beyond Social Security (GAO 2019b). Lower income levels leave people at greater risk in old age, and women and people of color are at greater risk of *both* lower income levels and low, or no, access to private pensions (Brown et al. 2016).

Without multi-employer pensions (Taft–Hartley funds), many more would be at risk of poverty in retirement. Multi-employer funds have helped millions of construction workers, trucking workers, performing artists, and others achieve some level of security in old age. Currently, there are a reported 10.6 million members supported by approximately 1,250 reported plans. Without such plans, several smaller businesses in those types of industries would likely offer no pension funds to their employees. The funds allow businesses to access trained labor who have long-term incentives to stay in those sectors.

However, many multi-employer plans are a threatened species. The financial situation for multi's before the Covid-19 pandemic was precarious for many, and now a greater number of already troubled plans face serious funding problems and risks of insolvency. The better funded multi's had just been bouncing back from dire predictions after the 2008 crisis, but those marked as critical and declining continued their long-term descent (Coffing et al. 2020). The reasons for this are many: fewer participants, accelerating plan maturity, employer withdrawals, long funding time frames, and risky investment practices. Decline in unionization rates worsens the issue, and a few very large funds had already accounted for the majority of the problem—for instance, the Teamsters' Central States Pension Fund (CSPF), which had 385,000 participants in 2019 and is projected to become insolvent by 2024 (GAO 2018). Others funds are smaller but were also troubled, such as the American Federation of Musicians and Employers' Pension Fund, which was the subject of the longest strike ever for the Chicago Symphony, last year. The strike was finally resolved when the players' union agreed to switch new employees to a defined-contribution plan because the orchestra was paying a great deal more to increase funding to the plan (Cooper 2019). As employers leave the plans due to much higher federal funding requirements, the multi-employer model is threatened. Even independently of the CSPF, the solvency problems of multi-employer plans will also exhaust the Pension Benefit Guaranty Corporation (PBGC) Multi-Employer Insurance Program, the program charged with providing partial benefit insurance in the event of a multi-employer plan insolvency. PBGC announced in November 2019 that it would be insolvent by 2025 if Congress does not act.

MEPs, another type of pension plan, can assist small employers in offering pension plans to their workers. While these plans allow small employers to pool their retirement plans while acting as a single entity, they are most often found in non-union settings where there is a common interest in business activities or associations. Individual employers do not share legal control, and the majority are defined-contribution plans resembling 401(k) plans (GAO 2012a). Contributions are made to a third party that is now considered the employer under the Employee Retirement Income Security Act of 1974, but the funds are not insured by the PBGC. The fiduciary liability facing employers is less than that for Taft–Hartley funds. Currently, the federal government collects little information about MEPs, and we know even less about the number of employers participating in a MEP because Form 5500s no longer contain information about employer contributions (GAO 2012b).

Both multi's and MEPs allow for some portability, but only among other employers within each plan, which tend to be along regional and sectoral lines. Individuals in MEPs who change employers may need to re-enroll and meet vesting requirements. However, depending on the MEP rules, employers may be able to spin off from the MEP, and individuals can move their accounts to new employers with similar 401(k) plans or to individual retirement accounts (IRAs). However, many employees are still tempted to cash out their distributions before retirement. Abandonment of plans upon separation are also a problem because many employees automatically enroll employees, and employees may not leave their forwarding addresses or the employer could go out of business. (GAO 2019a)

Another proposal for retirement accounts would be Teresa Ghilarducci's proposed guaranteed retirement accounts (GRAs). This plan, if approved, would follow employees from job to job, much like savings for colleges through 529 plans but administered by the Social Security Administration. The administration would not allow for early withdrawal of funds. A number of states have begun to pass state-based retirement accounts, but these plans can pose problems if people move and cannot avoid the early withdrawal problems of IRAs. GRAs have many admirable features to solve the retirement crisis, including hybrid (defined benefit/defined contribution) stability and independent oversight, but the problem seems to be one of political will to develop another federally based retirement system. In addition, the existence of GRAs could potentially undermine the political power of any remaining and well-functioning multi-employer pension plans.

Strategically, this becomes an issue for unions, their members, and for employers facing volatile production hours who want to offer employee benefits. Will they support federal or enhancement of state-based retirement accounts, or will they try to rescue the idea of the multi-employer pension funds or potentially develop their own retirement accounts? Some unions will want to try to control their own funds in cooperation with unionized employers and ensure those funds are stabilized, and others will look to the government to mandate greater retirement savings, usually through employer mandates or through general funds. These divisions among unions are not unlike other disagreements in the past.

Health and Welfare Funds

In 2018, the Commerce Department reported that 8.5 percent of Americans were still without health insurance at any point in the year. More than half (55 percent) of the population had employment-based benefits through their employer or union, and the percentage of employees offering health insurance is in decline (Berchik et al. 2018).

Multi-employer health and welfare funds are also allowable under Taft–Hartley provisions. These collectively bargained funds cover a host of medical, dental, visual, psychiatric, long-term care, severance benefits, accidental death or dismemberment, benefits, life insurance, and even unemployment, disability, or vacations or holiday benefits. In addition, some provide tuition assistance, daycare, and other subsidies. Like pensions, these can be either define benefit or defined contribution. In many cases, these funds go above and beyond what might be offered by the employer or through state-based health care programs. The added advantage is that these benefits are also portable within the sector of participating and unionized employers.

There are basically two methods foe funding health and welfare funds. Similar to pensions, they are either defined benefit for the cost of services with the level of payment based on claims, hours worked, or other factors. Or they are defined-contribution plans that are required to maintain individual accounts for each plan participant with payment limited to the amount contributed or earned on the individual's account.

Contributions could be voluntary or required under a collective bargaining agreement, and they can cover a single employer or multiple employers and participants. Or they can be noncontributory— meaning that only employers contribute.

In general, Taft–Hartley health and welfare funds covered a reported 4.9 million members, the majority of whom are active. The funds are fairly well capitalized with $58.75 billion dollars in income in 2017 and most plans meeting or exceeding annual expenses (Stoddard 2018). Additionally, they have an average overhead cost of that is fairly low compared with other healthcare funds. While some funds, particularly smaller funds, encounter variability in experiences, fund managers are advised to consider the timing of collective bargaining agreements and the ability to obtain changes in managing income and expenses. Again, some of these funds are quite large, with 35 funds covering more than 20,000 members.

Taft–Hartley health and welfare funds are facing significant challenges, however. The extension of the Affordable Care Act (ACA) has made the administration of these "Cadillac funds" more difficult. Some joint ventures are therefore looking to contain costs, add stop-loss coverage, or merge to capitalize on economies of scale. Several union members report that it has become more difficult to keep employers on board because they can encourage workers to seek public assistance (Hoffa et al. 2014). In addition, expansion of the Consolidated Omnibus Budget Reconciliation Act (COBRA) and other costs related to COVID-19 benefit extensions will place significant pressure on these funds. Again, as with pensions, laudable changes in the coverage of state-provided benefits may be undermining the Taft–Hartley plans' significant advantages for workers who change jobs often or experience breaks in employment but who stay within the same industry or occupation.

Hiring Halls

Another aspect of the multi-employer organizing strategy is control of the labor supply through a hiring hall. Hiring halls fall into two categories: exclusive and unexclusive. An exclusive hiring hall has collective bargaining agreements with employers who are able to call the hiring hall for workers. An unexclusive hall charges a small fee for a nonsignatory contractor to use the union hiring hall, and this can be a way for labor unions to have greater control of labor supply and quality even in right-to-work states. In addition to apprenticeship, the hiring hall becomes a means of preserving occupational standards. If a worker is not performing well, or engages in conduct unbecoming to the union, the worker may face sanctions from the union or employer, typically handled informally or grieved through its dispute resolution mechanism. The union leadership also may make special arrangements, particularly in the case of project labor agreements, to allow non-union community members to serve on large jobs of importance, such as a stadium or airport, with a view toward enrolling those workers in pre-apprenticeship or apprenticeship programs. One of the functions of a hiring hall, even one for day laborers, is to set the floor on wages and working conditions; therefore, they are not necessarily antithetical to union objectives (Theodore 2020). However, there are also drawbacks to hiring halls.

Because the hiring hall directs and assigns work, or if the employer relegates assignment of work to the hiring hall, the hiring hall legally becomes an employer and must comply with anti-discrimination and anti-retaliation laws while attempting to meet workers' demands for hours. Some hiring halls have become more fluid and informal, and others offer highly organized online hiring halls. Not surprisingly, claims of discrimination are common. The hiring hall determines who goes to which jobs. There is not a well-established set of rules for how a hiring hall ought to operate, and many locals have their own rules, although there is a body of legal precedent and best practices available through the Association for Union Democracy. In one case of many like it, *Daniels v. Pipefitters Association Local Union 597*, the court's remedy for claims of racial discrimination of an informal hiring hall was to place more rules regarding the local union's procedures: employers must notify the union of all jobs and use the hiring hall exclusively, qualified persons are to be referred in the order in which they signed the list, an orderly system of recording employment requests should be maintained, and union members have a right to review all records. Also contested was an employer's "right to reject" a union member. Some hiring halls will accede to the wishes of the employer to reject certain union members, thereby being complicit in employer-directed discrimination.

New types of hiring halls are emerging all the time—for example, medical professionals or college faculty who are nearing or at retirement and wish to earn a little bit more income before they reduce hours entirely. These employment leasing companies do not subscribe to union rules, of course, and many are considered managers who do not fall under the purview of the NLRA, but they nonetheless support employment for individuals who in many cases abide by a certain code of ethics, licensing, or accreditation standards.

Hiring halls can become political weapons, but they can be marshaled to provide well-trained workers in a pinch to employers who need them on short notice under certain employment conditions as well. Labor unions should therefore consider a few issues when establishing hiring halls. First, as with any multi-employer arrangement, there should be rules on entry and exit, and the local (or worker center) should have the capacity to handle calls and ration jobs as necessary. Because the union is under a different set of legal obligations than an employer or employment leasing company, there should be careful consideration and close relationships between hiring halls and their dispute resolution methods. The challenge would be to centralize these objectives while creating wage floors that are responsive to local labor market and living conditions.

Finally, to compete with the likes of hiring apps such as Uber, Lyft, and TaskRabbit, unions and signatory contractors need to create more active solutions to increasing demand for union trained labor. The technology on this is evolving but hiring-hall apps are in fact being developed.

Apprenticeship and Training

Quality of work is of paramount importance. Regardless of the hiring hall operation, the power comes from the strength of apprenticeship and training programs. Most training in formal apprenticeship programs are in the building and construction trades, although there are a growing number of programs in computer programming and in more traditional areas such as automotive repair and manufacturing; training programs for health aides and nursing assistants are also becoming more common. The vast majority of apprenticeship programs in construction sectors are provided by joint labor–management apprenticeship programs. Bruno and Manzo (2016) found that 98.5 percent of construction apprentices in Illinois in 2015 were enrolled in joint programs by union contractors, and program expenditures were nearly twice as great per apprentice in union programs than in non-union programs. These programs mainly serve those who fully commit to one occupation and are willing to make long-term career investments, and they are an important part not only for skills acquisitions but also education about the history and policies of their organizations. The opportunity for apprenticeship in other "apprenticable" jobs could expand to hundreds of other occupations according to the U.S. Department of Labor (Lerman 2019). This represents many opportunities for labor unions and for the workers they represent, as well as for companies who would employ them. It's a tremendous opportunity to address well=known free-rider effects, getting responsible employers to pressure their industry peers to promote a well-trained workforce.

Other than union-based training and registered apprenticeship programs, there are community college programs specializing in training for vocational and technical careers. Selected occupational programs in health care, information technology, and high-demand occupations have also yielded higher earnings.

Focused sectoral training programs such as Year Up and Work Advance have strong results (Fein and Hamadyk 2018; Hendra et al. 2016). However, in community colleges, graduation rates are generally low, and Pell Grants are limited, meaning that vocational training often gets students to only the first rung of the occupational ladder.

Training programs also provide ongoing skills and knowledge enhancement for people who continue in their trades. For example, OSHA 10-hour and 30-hour training programs are often offered by joint labor–management training centers. Training organizations' partners include union and non-union partners alike. New technologies also provide opportunities for training. For example, the United Brotherhood of Carpenters Pacific Northwest Carpenters Institute provides training on use of Bluebeam Revu, a cloud-based software program for digital measurement and mark-up of plans. Having workers trained in the software used by architects and engineers helps to improve productivity.

The many positive benefits of apprenticeship and training are beginning to expand, but getting employers to invest without the encouragement of unions or without collective bargaining agreements to finance such investments is challenging (Prebil 2020). As mentioned, the formal apprenticeship programs are largely the domain of the unionized building and construction trades who finance programs through cents per hour contributions and donations, and as such the programs tend to underrepresent residential workers who are more likely to be people of color, immigrants, and in contingent work situations. One proposal has been to provide training (or retraining) accounts to individuals. Another issue is to invest in childcare and transportation to assist low-wage workers who would be interested in completing training programs (Lerman 2019).

Dispute Resolution

Dispute resolution is also a vitally important part of managing employee–employer relationships in the multi-employer context. Here the construction industry has an enduring model, and many other building trades organizations have emulated it. This model is the IBEW-NECA joint labor–management partnership known as the Council on Industrial Relations for the Construction Industry. This remarkable board has been operating for just 100 years, predating the Wagner Act of 1935 and the Taft–Hartley Act of 1947. The Boilermakers and other unions have replicated the council, but more research needs to be performed on the effectiveness of these councils and their applicability to other sectors. It could be that the licensing and skills requirements of certain professions within construction require greater attention to the quality and skills development in those trades.

The subject of union involvement in dispute resolution is always controversial because many industrial unions have traditionally not wanted to give up the right to strike to resolve disputes. The reason the IBEW did so was multi-layered—essentially because they were not included in many war labor contracts but wielded considerable power owing to their geographical importance, high level of demand from consumers, and high level of skill that was required to wire buildings correctly and therefore keep electric grids humming. The contracting industry was facing tremendous pressure from large building companies to subsume electrical workers into their ranks and consequently (it was felt, by the National Electrical Contractors and Dealers Association at the time) to offer lower-quality services, while the contractors wanted to be able to move from job to job with their high-quality work and a highly trained workforce. They had to be able to keep those highly trained workers in the industry and provide a way to resolve any disputes quickly. The actual physical nature of the tasks, the geography, and the threat of substandard contracting practices were important to the development of structures that would promote the industry generally.

Evaluating the effectiveness of dispute resolution is challenging because the "success" of such outcomes is disputed and the details of arbitrated or mediated cases are often closed or available only until and when they are settled. Zach (2019) and Kochan (2019) stress that the decline in collective bargaining coverage produced a decline in dispute resolution systems that meet standards of due process, while more (about half) employees in non-union employment relationships are in lower-quality mandatory arbitration systems "imposed, designed and administered by employers." Decline in union density has encouraged widespread interest in smaller units of production and alternative dispute resolution, and this may be taking the place of collective, larger-scale sectoral expression of industrial–labor conflict. Hebdon, Douglas, and Mazerolle (1999) analyzed case studies from several industries and found that smaller bargaining units in Ontario were less likely to reach impasse. The researchers found evidence for a shift to more informal and individual expression of conflict and a shift away from collective expression. As Hebdon et al. emphasize, this does not mean that workplace conflict is in decline. In approximately 30 percent of arbitration cases, disputants went to courts; however, the arbitration awards were upheld 80 percent of the time.

Smaller bargaining units with more defined issues may not necessarily be antithetical to union goals, but to win the employers' agreement to participate, unions would have to show what the value added is to *employers* as well as to workers. Requiring transparency to the process may assist those in non-union dispute resolution mechanisms, but unions with well-functioning but somewhat opaque dispute resolution decisions would have to agree to this strategy as well.

Questions of Viability Remain

Over a century ago, many industry and labor leaders fought over the spoils of production, but some craft and "business" unions and trade associations agreed that labor–management systems could work, and they convinced other employers and local unions of the benefits toward the greater good. Now there are more complex legal, global, financial, and technological pressures at work, but a few sectors that rely on domestic labor could do more to bolster multi-employer approaches to industry problems. The need to craft effective institutions that serve the greater public good is no more apparent than it is now.

Organizing jointly managed union and multi-employer systems is difficult, and these systems are also challenging to administer. But importantly, they are still legal and function together as a labor and social compact to some degree. Standardized and moderate wage structure coupled with well-trained workers who add value could help to stabilize prices and distribute profits more evenly and to reduce turnover and waste. Dispute resolution with due process can maintain standards for workers and employers alike, as well as serve the public good. The internal mechanisms support the larger governmental goals of training and productivity. The advantage to a jointly managed system for employers is that they also can insist on high-road strategies as conditions of contracts affecting all organized employers.

It is important to note that these are systems that work together as economies of scope, potentially improving communication between skills demand and acquisition, increasing the tenure and craft identity of a proud and well-trained workforce, and resolving problems quickly and cost effectively. The incentives to develop and maintain these systems are partly economic and won't occur until the value of and the responsibility to the workforce is seen by employers and employer associations. Good hours, wages, safety, benefits, training, and resolution of disputes can help both workers and employers if there is some commitment to sharing the profits. Because many of these systems serve smaller employers and reduce administrative costs by pooling resources, there are also economies of scale. But many in government do not understand multi-employer arrangements, and their computer systems are not designed to adjust to alternative or multi-employer systems. Academia also tends to see multi-employer issues in the United States as mainly the domain of construction sector and white men engaged in business unionism. But it does work for many others, and perhaps we can imagine it for the socioeconomic support it can provide, at least theoretically. Further research is needed on experimentation in other sectors to be able to claim it will work in every sector. And the old multi-employer models also have significant financial problems to address, as I have noted.

To adopt multi-employer strategies, many unions also would have to develop a different approach. The advantage to unions in developing a multi-employer strategy is not just in obtaining benefits for members but in getting a chance to shape the conditions of work across all employers, not just individual employers. And they get to do this directly, as a condition of the contracts, because they control the supply of skilled labor, rather than having to rely on phone calls to overstretched regulators. To the degree that government fails to address full employment or portability of benefits, labor and management must look to their own solutions.

Finally, some serious questions need to be asked about what the degree of union instrumentality is, and what workers expect their unions to provide. Kochan (2019) reported from his worker surveys that workers "do not want to choose between the entrenched positions of labor or business. They want to have access to collective bargaining and informal participation processes as well as more individual career services and a voice in the governance processes of the firms that employ them." A renewal of multi-employer systems, with joint interests for workers and their unions, to employers and the public, is one time-tested way to accomplish this.

Acknowledgments

I am grateful to Charles Jeszeck of the U.S. General Accountability Office for his decades of guidance and prior collaboration on multi- and multiple employer pensions scholarship, and to Matthew Bodah of the University of Rhode Island for his collaboration on research that informed the dispute resolution section of this paper. All statements and any errors in this publication remain my own.

References

Beckhusen, Julia. 2019. Multiple Jobholders in the United States: 2013 Current Population Reports, P70BR-163, U.S. Census Bureau, Washington, DC: GPO.

Berchick, Edward R., Jessica C. Barnett, and Rachel D. Upton. 2018. Current Population Reports, P60-267(RV), Health Insurance Coverage in the United States: 2018, U.S. Census Bureau, Washington, DC: GPO.

Brown, Jennifer E., Joelle Saad-Lessler, Nari Rhee and Diane Oakley. 2016. Shortchanged in Retirement: Continuing Challenges to Women's Financial Future. National Institute on Retirement Security: Washington DC, March 2016.

Bruno, Robert and Frank Manzo. 2020. The Impact of Apprenticeship Programs in Illinois: An Analysis of Economic and Social Effects. University of Illinois—Champaign Urbana Labor Education Program August 24, 2016. https://bit.ly/3i4S7oY

Coffing, Kelly S. Timothy L. Connor and Nina M. Lantz. 2020. "Multiemployer Review: Covid 19 to leave multiemployer pension system more distressed than ever," Milliman, Inc. Reports April 2020. https://bit.ly/2GbFPhu

Cooper, Michael. 2019. "Chicago Symphony Ends Its Longest Strike With Pension Change," *New York Times*. April 27, 2019. https://nyti.ms/332QZ0M

Daniels v. Pipefitters' Assn., Local Union 597, 1990 U.S. Dist. LEXIS 12171, 135 L.R.R.M. 2534, 53 Fair Empl. Prac. Cas. (BNA) 1669 (United States District Court for the Northern District of Illinois, Eastern Division September 14, 1990, Docketed).

Fein, D. and Hamadyk, J. 2018. Bridging the Opportunity Divide for Low-Income Youth: Implementation and Early Impacts of the Year Up Program. OPRE Report #2018–65, Washington, DC: Office of Planning, Research, and Evaluation, Administration for Children and Families, U.S. Department of Health and Human Services. Washington, DC: GPO

Hendra, R., D. Greenberg, G. Hamilton, A. Oppenheim, A. Pennington, K. Schaberg, and B. Tessler. 2016. "Encouraging Evidence on a Sector-Focused Advancement Strategy: Two-Year Impacts from the WorkAdvance Demonstration." New York, NY: MDRC

Hoffa, Geoffrey D., Kathleen Mathieson, Catherine Belden, and Simone Rockstroh. 2014. "Taft Hartley Health Funds face many challenges including ACA," *Benefits Magazine* July: p.48-53.

Kochan, Thomas A., and William T. Kimball. 2019. "Unions, Worker Voice, and Management Practices: Implications for a High-Productivity, High-Wage Economy." *RSF: The Russell Sage Foundation Journal of the Social Sciences* 5(5): 88–108. https://bit.ly/337jj2l

OECD. 2018. "OECD Employment Outlook 2018," OECD Publishing, Paris, France: Organisation for Economic Cooperation and Development. https://bit.ly/3cFiyAz

Lerman, Robert, Pam Pamela Loprest Daniel Kuehn. 2019. "Training for Jobs of the Future Improving Access, Certifying Skills, and Expanding Apprenticeship." Oct. 3. Washington, DC: Urban Institute. https://urbn.is/2Gc9Q0F

Manyika, J., S. Lund, M. Chui, J. Bughin, J. Woetzel, P. Batra, R. Ko, and S. Sanghvi. 2017. "Jobs Lost, Jobs Gained: What the Future of Work Will Mean for Jobs, Skills, and Wages." Washington, DC: McKinsey Global Institute.

Prebil, Michael. 2020. "Tech Apprenticeship in the San Francisco Bay Area." March 2020. *New America.* https://bit.ly/3cG3Ppa

Stoddard, David. 2019. "Multiemployer health and welfare fund statistics: Fall 2019," Nov. 1. Milliman White Paper. https://bit.ly/335jGKM

Theodore N. 2020. "Regulating informality: Worker centers and collective action in day-labor markets." *Growth and Change* 51:144–160. https://bit.ly/3j6GdMK

U. S. Government Accountability Office, 2009. Employee Misclassification: Improved Coordination, Outreach, and Targeting Could Better Ensure Detection and Prevention GAO-09-717. August 10, Washington, DC: GPO.

U.S. Government Accountability Office. 2012a. Private Pensions: Better Agency Coordination Could Help Small Employers Address Challenges to Plan Sponsorship. GAO-12-326, March 5. Washington, DC: GPO.

U.S. Government Accountability Office. 2012b. Private Sector Pensions: Federal Agencies Should Collect Data and Coordinate Oversight of Multiple Employer Plans. GAO -12-665. Sep. 13. Washington, DC: GPO.

U.S. Government Accountability Office. 2015. Contingent Workforce: Size, Characteristics, Earnings, and Benefits. GAO 15-168R. Apr 20. Washington, DC: GPO.

U.S. Government Accountability Office. 2018. Central States Pension Fund: Department of Labor Activities under the Consent Decree and Federal Law. GAO-18-105 Jun 4. Washington, DC: GPO.

U.S. Government Accountability Office. 2019a. Retirement Accounts: Federal Action Needed to Clarify Tax Treatment of Unclaimed 401(k) Plan Savings Transferred to States. GAO-19-88, Jan.18. Washington, DC: GPO.

U.S. Government Accountability Office. 2019b. Retirement Security: Most Households Approaching Retirement Have Low Savings, an Update. GAO-19-442, Jan 29. Washington, DC: GPO.

Weil, D. 2014. *The Fissured Workplace: Why Work Became So Bad for So Many and What Can Be Done About It*. Cambridge, MA, Harvard University Press.

Zack, A.M. 2019. "Reclaiming Arbitration's Good Name." *Dispute Resolution Journal* 74(1), 33-42.

III. LERA Best Posters Session

Open Channels: Workplace Dispute Resolution Systems for Sustainable Downsizing

COURTNEY CHICVAK
Columbia University

In a downsizing scenario, how can workplace dispute resolution systems be coupled with more traditional processes such as retraining, transfers, and notice periods for non-union employees to ensure long-term organizational sustainability? Considering both employee and employer perspectives, this poster proposes that opportunities exist to build long-term relationships with employees involved in downsizing through the implementation of a dispute resolution process that runs concurrently with more traditional processes.

Labor Market Explanations for the Decline of New England Whaling

NITI PANDEY
Eastern Connecticut State University

Whaling as an organized industrial enterprise peaked in the mid-19th century before seeing sustained decline, effectively coming to an end in the 1920s even as global demand for the products of whaling remained high. In spite of the size and impact of the industry on the economy of New England in the 18th and 19th centuries, very little research exists on the labor and business aspects of the industry. This research explores labor market factors such as the unique "lay" wage system, labor supply, competing labor markets, occupational restrictions, and global labor mobility as possible explanations for the industry's decline.

Craving More: The Influence of a Small Coffee Shop on Social Sustainability

ROB BOYLE
Saint Louis University

It has been argued that an increasingly social perspective on sustainability is often overshadowed by extremely important environmental concerns such as the development of eco-friendly buildings that minimize carbon footprints. More and more, however, there seems to be a strong movement toward balancing both the needs of the environment as well as the psychological needs of people.

It is, indeed, healthy people and livability that are at the heart of Crave Coffeehouse in St. Louis. Crave is the culmination of a vision shared by four people: a Lutheran pastor, a local seminary student, and two caring, resourceful, and innovative community members named Phil and Donna Green. The four agreed on the name Crave because it naturally alludes to the mission of providing outreach to people who crave peace, crave coffee, and crave more in life.

A lengthy interview with one of the visionaries, Donna Green, revealed that social sustainability is at the heart of Crave's day-to-day operations. Green was also the manager of Crave Coffeehouse for the first ten years of its existence, and as such she has colorful and important insights about the sophistication involved with running a sustainable social venture. She also understands the historical significance of the ministry and the legacies of the countless lives that are touched. She understands the potential beauty and draw of a building that was damaged by fire, covered in dust, and falling into disrepair. She understands the value of the place and the sense of peace, calm, and safety that it offers to employees and those who come through the doors. And she understands the value of the surrounding community that needed a positive and productive small business to bring hope to an ailing neighborhood.

IV. LERA Competitive Papers II: Labor Unions and Employee Voice

Firms as Targets of Union Pension Shareholder Activism

BRANDON CARLYLE GRANT
SUNY–Farmingdale State College

Introduction

Activists in social movements have a long-standing tradition of targeting firms for the purposes of eliciting corporate change. From traditional methods such as protests, boycotts, and peaceful sit-ins to contemporary methods such as online reviews, blogs, and social media campaigns, social movements have no shortage of means by which to have their message heard by the media, the general public, and of course, the target firm. While the medium continues to evolve, the spirit of these movements remains largely unchanged; the goal of social movements is change. Yet until recently, scholars studying social movements had mostly overlooked the process by which firms came to be targets for change in the first place.

Bartley and Child (2014) provide strong empirical support for their argument that firm size, strength, and network position/affiliation are significant predictors of whether a firm will be a target for activism. This paper examines the characteristics of firms and firm reputation for downsizing as antecedents to being targeted for shareholder activism by labor unions. Firm characteristics are important for predicting many outcomes, including stock price (Rezee, Espahbodi, Espahbodi, and Espahbodi 2012), targets of activism (Soule 2009), and performance (Hendricks and Singhal 2001). The relationship between a union and management is important for determining future outcomes that affect both parties, and when it comes to firm performance and shareholder value, both parties have a stake in seeing a firm do well.

From the perspective of a union, they can find themselves in a very precarious position as both the workers for and partial owners of the firms that they represent. This puts them in the unique position of being able to potentially affect corporate change from multiple perspectives simultaneously. Whether it be in the voting booth during a certification election, the bargaining table during collective bargaining agreement negotiations, the picket line during a strike, or the annual meeting during shareholder proxy proposal discussion and voting, unions have many points of opportunity to voice their concerns about issues regarding management policies and decision making. Depending on the relationship that is maintained between the two parties (union and management), these procedures can all go a number of ways, from peaceful and professional to contentious and aggressive (and anywhere in between). The research question in this paper is what factors cause some firms to be targeted for union shareholder activism and not others?

To begin, this paper will examine the ever-evolving nature of the relationship between labor unions and management. Historically, unions and management have tended to have a rather contentious relationship. While this is not always the case, it is considered by most to be the norm. Having said that, over time we have seen an evolution in this relationship, and many unions and firms have begun to display more cooperative relationships. For example, in their 2015 book, Cutcher-Gershenfeld, Brooks, and Mulloy discuss the transformative nature of the UAW–Ford transformation that took place during a time of major economic crisis in the United States, allowing Ford to emerge as the only major auto manufacturer in the country that did not require a government bailout to remain afloat. As the nature of union–management relationships continues to evolve, you might expect to observe evidentiary outcomes that align with the nature of said relationship. This paper seeks to show evidence that certain union characteristics, as well as firm characteristics and the nature of the relationship between unions and firms, are useful predictors in determining whether or not a firm will be targeted for shareholder activism.

I look at firm governance indicators such as financial performance and employment levels as predictors of being targeted by labor union shareholder activists. Each of the key independent variables provides important potential explanatory power in predicting the probability that a firm will be targeted by union shareholder activism. Firm characteristics are important to the extent that I expect that firm size and firm finances, as well as firm industries and employee relations, are related to the dependent variables of interest. Perhaps of greatest significance is the nature of the firm's reputation for downsizing. Using firm employment level data as a proxy for the downsizing, I expect to find that firms involved in downsizing will be more likely to be targeted by union shareholder activism. Additionally, I expect to find that as firms lay off larger numbers of workers, they will be increasingly more likely to be targeted for shareholder activism by unions.

Consistent with the old adage, "the squeaky wheel gets the grease," one should expect that firms with greater reputations for downsizing will be more likely targets for union shareholder activism. Bartley and Child (2014) write extensively on how certain corporations come to be targeted by corporate campaigns and other social movements. Those authors show that certain types of firms with certain characteristics are more likely to be targets of social movements. In addition to power and network position, the authors show that a firm's relationship with various stakeholders is a powerful predictor of the probability that they will be targets of social movements. Building on this and similar research, this paper looks to extend the theory of Bartley and Child (2014) by demonstrating that, in addition to firm and union characteristics, the nature of the relationship between a union and a firm will be a strong and significant predictor of the probability that a firm will be targeted for shareholder activism by unions.

In addition to the probability of being targeted for shareholder activism, this paper also examines the frequency with which a firm is targeted by unions as activist shareholders. I argue that the simple probability of being targeted for shareholder activism opens the door for this second set of statistical models, which build on the first models and the first dependent variable. I argue that in instances when firms are targeted repeatedly, this offers additional insight beyond what the first set of models reveals—that unions use their roles as shareholder activists as a means of signaling to firms that their employment relations policies and practices do not go unmonitored.

Literature Review

An increasingly common phenomenon in the sphere of social movements is the targeting of large, visible firms for the alleged purpose of eliciting social change (i.e., Bartley and Child 2007; King and Soule 2007; Soule 2009; Vasi and King 2012). Institutional investor shareholder activism is a particularly popular and growing topic within the scholarly literature on social movements (i.e., Gillan and Starks 1998, 2000, 2007). In their meta-analysis on corporate governance, Daily, Dalton, and Cannella (2003) pointed out the paucity of research in the area of shareholder activism as a key form of corporate governance, and scholars eagerly responded to their call for research in this area over the ensuing 15 years. Westphal and Bednar found evidence that CEOs commonly use their positions and power to pacify investors through "ingratiation and persuasion" in order to deter them from attacking them and their firms through the proxy proposal process (2008: 29). As the findings of the related research shows, unions are capable of being coerced into withdrawing shareholder resolutions, just as other institutional investors are. In a study of shareholder activism and corporate social performance, David, Bloom, and Hillman (2007) found that firms divert funds from socially responsible practices in order to increase resources available for protecting themselves from external attacks such as shareholder activism.

In addition to these and many other studies, some scholars have documented anecdotal evidence of firm responses to shareholder activism. Gillan and Starks (2007) discuss the role of Fidelity, one of the largest institutional investing firms in the United States, in the eventual departure of then-CEO Kay Whitmore from Eastman Kodak. Webber (2018) gives a heavily detailed account of how the California Public Employees' Retirement System led the charge in ousting Michael Eisner from his seat atop the Disney empire, as well as how the American Federation of Teachers (AFT) has gained a reputation in investing communities and circles for its annual publication of the "Investment Manager Watch List" by naming and shaming hedge funds and their managers for their irresponsible investing practices. As Webber chronicles, Randi Weingarten and the AFT's annual watch list has so much clout that top hedge fund managers have been known to personally reach out to Weingarten (president of the AFT) in order to negotiate on how to be removed from the list.

One important thing that all of the aforementioned studies have in common: They all study firm responses as outcomes of shareholder activism. Recently, the most common criticism of scholarly research on shareholder activism is that despite the rapid and vast growth in the research on shareholder activism and social movements, there exists a dearth of research that focuses on the antecedents of shareholder activism, and what research does exist on the predictors of shareholder activism is largely ambiguous.

Goranova, Abouk, Nystrom, and Soofi (2015) conducted a study in which they decouple two important and distinct latent constructs that they show to separately affect shareholder activism: shareholder activist characteristics and firm characteristics. Specifically, they show that shareholders who are more dissatisfied with firm performance may be more likely to engage in shareholder activism, while firms might be more or less likely to acquiesce to the demands of shareholder activists, and that corporate governance plays a role in mediating the causal relationships between these antecedents and shareholder activism outcomes.

"The most consistently tested drivers of shareholder activism are, by far, firm size and performance" (Goranova and Ryan 2014: 1242). As these authors point out, the vast majority of research that does empirically analyze the predictors of shareholder activism look primarily at firm characteristics. Commonly relying on agency theory, research on predictors of shareholder activism logically argue that poor-performing firms *should* be common targets of shareholder activism because shareholders have a financial stake in the firm's performance. Bizjak and Marquette (1998) and Prevost and Rao (2000) argue that firms with higher proportions of managerial ownership are less likely to be targets of shareholder activism because management, as shareholders, has a financial stake in their own firm and thus their interests are aligned with other shareholders.

In addition to target firms, shareholder activists have also been examined as antecedents to shareholder activism, but to a lesser extent than firms themselves. According to Goranova and Ryan, "focusing on target firms alone without considering the interests, identity, concerns, and considerations of the activists could paint a partial picture of shareholder activism, at best, and a misleading one, at worst" (2014: 1244). What these authors are so shrewdly pointing out is that understanding activism in terms of targets without considering anything about those engaging in the actual behavior makes little sense. The most logical explanation for the paucity of research on activists as antecedents might simply be lack of accessible data on activists. According to Benton (2017), the largest group of shareholder activists is individual investors, the second largest is unions, and the third largest is institutional investor organizations/groups. Individual investors are private citizens, and institutional investing groups are (typically) private firms, as such public accessibility to financial information and other data is limited, if not nonexistent. However, under the Labor–Management Reporting and Disclosure Act of 1959, labor unions in the United States are required to file annual financial disclosure forms with the US Department of Labor, and thus their data are publicly available.

This makes labor unions a lucrative group for investigating the area of activists as the level of analysis, particularly to the extent that they bear a dual-agency role to the constituencies they represent (financial agents and bargaining agents).

When studying the activists as the level of analysis, it is important to understand that shareholders come in all cuts and colors. Some institutional investors such as unions, churches, and nongovernmental organizations (NGOs) are nonprofit in nature and as such bear a dual-agency role to their constituents that goes beyond a singular profit-maximization role. However, unlike churches and NGOs, unions exist as corporate shareholders through their pension funds, which must earn money to remain sustainable over time. Thus, unions potentially bear a different cost of activism than other nonprofit organizations, and they certainly have a cost of activism than differs from that of traditional for-profit institutional investor groups. Costs of activism can vary depending on the type of institutional investors and, as such, the amount of stake (financial or otherwise) that an investor has in a firm can vary as well (Cuñat, Gine, and Guadalupe 2012).

Borrowing from the line of reason of Bizjak and Marquette (1998), Prevost and Rao (2000), and others, this study relies on agency theory to explain the hypotheses and results. The purpose of this study is to show that when firms are larger, more financially stable, and have more contentious relationships with employees (i.e., more downsizing), union-sponsored shareholder resolutions will be more likely and in greater quantity because as bargaining agents on behalf of workers, unions have a principal-agent responsibility to workers that goes beyond simple financial interests.

Theory and Hypotheses

The literature on shareholder activism has established that firms that are larger and more visible are more likely to be targets of shareholder activism for a variety of reasons. Unions as shareholders through their pension funds have a dual-agency role to their constituents such that they must not only maximize profit to the extent that it prolongs the life of the pension fund for the foreseeable future, but unions also exist as the bargaining agents of their workers in the workplace. The hypotheses in this study reflect this dual-agency role.

In addition to controlling for firm size and financial performance, all models include industry controls, year-effects controls, and CEO Sarbanes–Oxley (SOX) certification controls. Industry controls are important to rule out any variance in the dependent variables that are industry specific; it could be that some industries are more prone to being targets of shareholder activism for any number of reasons. Annual controls are important to the extent that over time trends might change; for instance, following the major economic recession in and around 2007–2008, monitoring of corporate governance peaked as many Americans became increasingly concerned about the values of their investment portfolios. SOX certification is a program that awards certifications to professionals for undergoing a training program through the Governance, Risk, and Compliance (GRC) Group and is a commonly earned certificate for upper-level managers and chief officers at the S&P 1500. SOX certification is a signal to others—including directors and stockholders—that the firm seeks to hold itself to higher compliance levels and standards. Ideally, by controlling for SOX certification I can rule out any omitted variable bias (OBV) associated with potential past actions or issues that would have caused a firm to pursue SOX certification in the first place (such as corporate scandals).

Larger firms are more likely to attract large investors, and as such, attract more scrutiny and governance monitoring. Union and non-union shareholder activists alike will have an increased probability and frequency of targeting large firms because they are likely to have a larger financial stake in larger firms. Larger firms therefore would be expected not only to be more likely targets of shareholder activism but more *frequent* targets because of this increased level of monitoring. Therefore,

- H1-1: Firm size will positively predict the probability of being a target of union-sponsored shareholder resolutions.
- H1-2: Firm size will positively predict the frequency of being a target of union-sponsored shareholder resolutions.

In addition to firm size, past literature on shareholder activism suggests that firm financial performance is a strong predictor of shareholder activism for reasons similar to size. Financially stable and strong-performing firms are more likely to attract larger investors and increased monitoring and scrutiny from investors. Thus,

- H2-1: Firm financial performance will positively predict the probability of being a target of union-sponsored shareholder activism.
- H2-2: Firm financial performance will positively predict the frequency of being a target of union-sponsored shareholder activism.

Controlling for standard firm characteristics (firm size and financial stability) as well as year and industry, I expect that unions will respond to two key types of firm changes in two distinct and different ways. Due to their dual-agency role, I expect that unions as shareholders will respond to firm financial changes similarly to traditional investors by increasing likelihood and frequency of shareholder activism. As such, I predict that when financial performance goes down, probability and frequency of being targeted by union shareholder activism will go up:

- H3-1: Firm decreases in financial performance from the previous year will positively predict the probability of being a target of union-sponsored shareholder activism.
- H3-2: Firm decreases in financial performance from the previous year will positively predict the frequency of being a target of union-sponsored shareholder activism.

In addition to their position as a financial agent on behalf of members, labor unions also play a servicing role for their members as their bargaining agent in the workplace. As H4-1 and H4-2 predict, the fiscal agency role predicts that when firm's financial performance changes, shareholder activism will too. On the other side of the dual-agency coin, unions have a responsibility to protect their members in the workplace. Therefore, I predict that when employment goes down (downsizing), the probability and frequency of being a target of union-sponsored shareholder activism will go up:

- H4-1: Firm changes in employment (downsizing) from the previous year will positively predict the probability of being a target of union-sponsored shareholder activism.
- H4-2: Firm changes in employment (downsizing) from the previous year will positively predict the frequency of being a target of union-sponsored shareholder activism.

Data and Methods

For this study, I used employment data from Compustat on S&P 1500 firms as a proxy for downsizing, as well as the financial data on firms to explain patterns of shareholder activism and combined downsizing data with the Institutional Shareholder Services (ISS) data on those same S&P 1500 firms. In many ways, this study builds on related research by responding to recent criticisms in the shareholder activism literature that research on the antecedents of shareholder activism is drastically underexamined. Here, I gain a more holistic understanding of the big picture of union shareholder activism. The summary statistics can be found in Table 1.

Both of the dependent variables in this study (likelihood and frequency of being targeted by union-sponsored shareholder proposals) come from the ISS database. The first dependent variable is a binary variable where 0 represents a firm that was not targeted by any union-sponsored shareholder proposals in a given year, and a 1 represents a firm that was a target in that year. The second dependent variable is a count variable that represents the number of times that a given firm was targeted by any union-sponsored shareholder resolutions in a given year.

The independent variables in this study come from the Compustat data repository, and all are logged unless otherwise specified. The key independent variables in this study are employee count (which I use as a proxy for firm size), earnings before interest and taxes (or EBIT, which I use as a proxy for firm financial performance), changes in employee levels (which I use as a proxy for downsizing), and changes in EBIT. I also include controls for year and industry. Industry codes are included as a set of dummy variables representing the S&P Industry Code (SPC) for each firm. SOX is a binary variable representing whether or not the CEO was, in that year, SOX-certified.

Table 1. Summary Statistics and Bivariate Correlations

Variable	Mean	S.D.	Min	Max	1	2	3	4	5	6
1. Resolution	0.026	0.158	0	1						
2. Resolution Count	0.042	0.295	0	7	0.87					
3. EBIT*	4.033	2.213	–6.908	11.395	0.26	0.24				
4. Employee Count*	–0.025	2.234	–6.908	7.842	0.23	0.21	0.72			
5. Employee Change*	0.018	0.29	–6.225	8.642	–0.01	–0.01	0.01	0.09		
6. EBIT Change*	0.044	0.519	–9.395	8.124	0	0	0.15	0.02	0.09	
7. CEO SOX Cert	0.66	0.474	0	1	0.02	0	0.1	0.04	0.07	0.06

* = Logged variable

The first set of models use random-effects logistic regression to test a binary dependent variable, which represents whether or not a S&P 1500 firm was targeted for union shareholder activism in a given year. I chose to use random effects for my first set of models because the firms in the dataset do not consistently appear in the data from one year to the next. While the appearances of firms in a given year is not truly random, it is not fixed enough to warrant the use of fixed-effects logistic regression or mixed-effects logistic regression. The unit of analysis in all models is the firm, and the dataset includes all S&P 1500 firms from the year 2000 to the year 2014. The second set of models use negative binomial regression to test the frequency with which firms by year are targeted by union shareholder activism. Because this is likely a highly skewed, overly dispersed distribution with a disproportionate number of zeros, negative binomial regression allows me to run the regression despite the mean of the distribution being so heavily skewed. These data merge data from ISS with data from the Compustat data repository using the Committee on Uniform Security Identification Procedures company identifiers, and then finally were collapsed by firm to create a unique set of data with the firms by year as the unit of analysis. The coding schemes and frequencies for the variables in my analyses can be found in Table 2.

Table 2. Frequencies

Dependent Variable	Coding Scheme	Frequency Info
1. Resolution	0 = no resolution submitted; 1 = resolution submitted	n = 2225
2. Resolution Count	0–7 range; continuous variable	n = 2227
Independent Variable	**Coding Scheme**	**Frequency Info**
3. EBIT*	Continuous, in millions	Mean = 425.42
4. Employee Count*	Continuous, in millions	Mean = 7.30
5. Employee Change*	Continuous change from previous year	Mean = .018
6. EBIT Change*	Continuous change from previous year	Mean = .044
7. CEO SOX Cert	0 = no SOX certification; 1 = SOX certification	Mean = .66

* = Logged variable

In testing hypotheses 1–4, the dependent variable was binary, representing whether a firm was or was not the target of union-sponsored shareholder activism. The data comes from Institutional Shareholder Services (ISS) and provides information on the names of the unions that submitted each proposal (which I used to match to the unions in my dataset), as well as the company for whom the resolution was submitted, the results of the resolution (withdrawal, voted on, passed vote, did not pass vote, etc.), and a company identifier that can be used to match companies to other datasets and databases such as Compustat. For these four hypotheses, I coded the dependent variable based on whether or not the firm was the target of at least one shareholder resolution in a given year. In all models, I controlled for year, industry, and CEO SOX certification. The four models are stepwise, testing each key independent variable, with the fourth model being the full model.

For hypotheses 5–8, the dependent variable is continuous and represents the number of times (ranging from 0 to 7) that a firm was targeted by shareholder activism in a given year. As was the case in testing the first four hypotheses, the data also comes from ISS and Compustat; however, given that the dependent variable is a discrete count variable with zero-inflation (approximately 97 percent of all 86,581 observations were zeros), I used negative binomial regression to test the remaining hypotheses and their accompanying models. As before, the models are stepwise, and the eighth is the full model discussed below. The hypotheses are summarized in Table 3.

Table 3. Hypothesis Statements and Results

Hypothesis Number	Hypothesis Statement	Hypothesized Direction	Actual Direction
H1-1	Firm size will positively predict the probability of being a target of union-sponsored shareholder activism.	+	+
H2-1	Firm financial performance will positively predict the probability of being a target of union-sponsored shareholder activism.	+	+
H3-1	Firm changes in financial performance from the previous year will negatively predict the probability of being a target of union-sponsored shareholder activism.	–	–

H4-1	Firm changes in employment (downsizing) from the previous year will negatively predict the probability of being a target of union-sponsored shareholder activism.	–	–
H1-2	Firm size will positively predict the frequency of being a target of union-sponsored shareholder activism.	+	+
H2-2	Firm financial performance will positively predict the frequency of being a target of union-sponsored shareholder activism.	+	+
H3-2	Firm changes in financial performance from the previous year will negatively predict the frequency of being a target of union-sponsored shareholder activism.	–	–
H4-2	Firm changes in employment (downsizing) from the previous year will negatively predict the frequency of being a target of union-sponsored shareholder activism.	–	–

Results, Discussion, and Conclusion

In the test for hypothesis 1-1, the results in Table 4 show that with each unit increase in logged employee count (a proxy for firm size), the likelihood of a firm being targeted by a union-sponsored shareholder resolution increases by a factor of .38 ($p < .01$). Column 4b of Table 4 shows the odds ratio for the full model, indicating that a unit increase in log employees means a firm is 45 percent more likely to be targeted for shareholder activism ($p < .01$). Model 8 shows the results of testing hypothesis 1-2, which predicts that firm size will be positively related to the frequency of being targeted by union-sponsored shareholder activism. The results show that a unit increase in log employees leads to a .38 increase in frequency of being targeted by union-sponsored shareholder activism.

Hypotheses 2-1 and 2-2 predict the likelihood and frequency of being targeted by union-sponsored shareholder activism by firm financial performance. The results of testing hypothesis 2 show a unit increase in log earnings before interest and taxes (EBIT) causes a 91 percent increase in the odds that a firm will be targeted by union-sponsored shareholder activism ($\beta = .65$, $p < .01$). Results of testing hypothesis 6 show that each unit increase in log EBIT means a .56 increase in the frequency that a firm is targeted by union-sponsored shareholder activism.

In a downsizing scenario, how can workplace dispute resolution systems be coupled with more traditional processes such as retraining, transfers, and notice periods for non-union employees to ensure long-term organizational sustainability? Considering both employee and employer perspectives, this paper proposes that opportunities exist to build long-term relationships with employees involved in downsizing through the implementation of a dispute resolution process that runs concurrently with more traditional processes.

According to hypotheses 3-1 and 3-2, decreases in employment rates at a firm (downsizing) should lead to an increase in both the probability and frequency of being targeted by union-sponsored shareholder activism. Results in Tables 4 and 5 show support for both hypotheses. Models 4 and 4b indicate that a firm is 49 percent more likely to be targeted by union-sponsored shareholder activism ($\beta = .68$, $p < .01$) with each unit decrease in log employment change. Model 8 shows that employee downsizing increases the frequency of being targeted for shareholder activism by a factor of $\beta = .65$ ($p < .01$) with each unit change.

Table 4. Models 1–5, Random-Effects Logistic Regression
(Dependent variable: Probability of firm being targeted by union-sponsored shareholder activism)

Models	**1**	**2**	**3**	**4**	**4b (OR)**
Independent Variables					
Employee Count	0.39*** (0.04)	0.40*** (0.04)	0.36*** (0.04)	0.38*** (0.04)	1.45*** (0.06)
EBIT	0.63*** (0.04)	0.62*** (0.04)	0.66*** (0.04)	0.65*** (0.04)	1.91*** (0.07)
Employee Downsizing		0.79*** (0.16)		0.68*** (0.16)	1.49*** (0.08)
EBIT Decrease			0.36*** (0.08)	0.29*** (0.08)	1.25*** (0.06)
CEO SOX Cert.	4.37*** (1.26)	4.39*** (1.27)	4.39*** (1.30)	4.40*** (1.30)	81.80*** (106.16)
Year					
	Inc.	Inc.	Inc.	Inc.	Inc.
Industry Code					
	Inc.	Inc.	Inc.	Inc.	Inc.
Constant	−13.53*** (1.06)	−13.54*** (1.06)	−13.64*** (1.06)	−13.55*** (1.06)	−1.30e-06*** (1.06)
Observations	45,641	45,641	45,641	45,641	45,641
Pseudo R-square	0.378	0.379	0.379	0.381	0.381

Standard errors in parentheses. ***p < 0.01, **p < 0.05, *p < 0.1

Table 5. Models 6–10, Negative Binomial Regression Models
(Dependent variable: Frequency of firm being targeted by union-sponsored shareholder activism)

Models	**5**	**6**	**7**	**8**
Independent Variables				
Employee Count	0.40*** (0.03)	0.41*** (0.03)	0.37*** (0.03)	0.38*** (0.03)
EBIT	0.56*** (0.03)	0.55*** (0.03)	0.59*** (0.03)	0.58*** (0.03)
Employee Downsizing		0.75*** (0.15)		0.65*** (0.15)
EBIT Decrease			0.32*** (0.07)	0.26*** (0.07)
CEO SOX Cert.	4.42*** (1.19)	4.44*** (1.20)	4.43*** (1.21)	4.44*** (1.21)
Year				
	Inc.	Inc.	Inc.	Inc.
Industry Code				
	Inc.	Inc.	Inc.	Inc.
Constant	−12.19*** (0.77)	−12.20*** (0.77)	−12.28*** (0.77)	−12.17*** (0.77)
Observations	46,402	46,402	46,402	46,402
Pseudo R-Square	0.307	0.308	0.308	0.312

Standard errors in parentheses. ***p < 0.01, **p < 0.05, *p < 0.1

Finally, hypotheses 4-1 and 4-2 predict that changes in firm performance will be negatively related to both the probability and frequency of a firm being targeted for shareholder activism. Results show that a unit change in log firm performance (drop in EBIT) leads to a 25 percent increase in likelihood of being targeted by union-sponsored shareholder activism ($\beta = .29$, $p < .01$) and an increase of $\beta = .26$ ($p < .01$) in the frequency that a firm will be targeted by union-sponsored shareholder activism.

Overall, the results show strong support for all hypotheses. First, this study replicates previous research and findings that show how larger and more financially successful firms are frequently the targets of shareholder activism for a variety of reasons. While controlling for these known firm-specific characteristics, the results shown here also indicate support for the hypothesis that firm downsizing strongly predicts the likelihood and frequency of retaliation by unions through the shareholder resolution process while they are simultaneously responsible to their constituents as financial agents. Taken together, these results support the dual-agency theory put forth here that unions play to their principle constituents—i.e., their members.

As Goranova and Ryan (2014) state, the overwhelming majority of research on the predictors of shareholder activism focuses on firm size and financial strength. While this study does follow that trend—and this could be argued to be a limitation of this study—the findings in this study also build further on the previous research in a couple of meaningful ways. The first contribution of this study is that it does not exclusively focus on firm size and financial strength, but instead controls for those criteria while adding an important and previously unexamined predictor variable—employment relations/downsizing. This is important when considering labor-union pension funds as the shareholder activists of interest because unions serve a dual role as both potential and/or current owners of publicly traded firms, as well as potential and/or current bargaining agents on behalf of workers at those same firms.

The second contribution of this study is that it opens the door for future research on other important characteristics about shareholders that could focus on the shareholders themselves as predictors of shareholder activism. Future research on the antecedents of labor-union pension shareholder activism should examine characteristics of the unions themselves and how they predict the probability and frequency of engaging in shareholder activism. A logical next step in the sequence of research in this study would be to extend it further by looking at union size, financial strength, and relationships/affiliations as predictors of shareholder activism. Labor unions are the second-largest institutional investor group in the United States, behind only individual investors (Benton 2017). As such, research such as this study can and does have a profound impact on the scholarly literature on corporate governance, shareholder activism, and labor relations. Corporate governance literature still lacks some depth in management journals because it pertains to shareholder activism, especially predictors of shareholder activism, and the empirical research on labor unions as shareholder activists is quite scant to date.

This study's third and final contribution is that it sets the table for future research on the dual-agency role that some institutional investors have that goes beyond the fiscal responsibility it has to the people that it represents. Not all institutional investing organizations necessarily serve their constituents exclusively as financial agents, and this could potentially carry with it important explanatory power for shareholder activism outcomes. This study shows evidence that firm downsizing may be related to shareholder activism such that labor unions appear to respond to downsizing with increased frequency and probability of targeting a given firm with one or more shareholder resolutions. While controlling for firm financials, the results still emerge statistically significant. This may fly in the face of conventional wisdom of past scholarly claims that institutional investors have a singular responsibility to their constituents to maximize shareholder value. As Chava et al. (2010), Kumar and Ramchand (2008), and Webber (2018) suggest, shareholders might occasionally pursue interests that are misaligned with other shareholders, and if that is the case it should follow that other shareholders would fail to support those resolutions in the event that they go to a vote.

Future research on shareholder activism should seek to unpack the contextual factors surrounding shareholder activism by institutional investor groups that might consider themselves to have a dual-agency role, such as unions, churches, and NGOs. Soule (2010) discusses the activist roles of churches and NGOs and how they seek to affect corporate governance through contentious activities such as social movements. Bartley and Child (2007) discuss the phenomenon of "naming and shaming" lists and how nongovernmental agencies can affect corporate change through public shaming. Similarly, as previously mentioned, Webber (2018) documents the narrative of hedge fund name-and-shame lists sponsored by the AFT. Extensive empirical research of these dual-agency institutional investor groups such as their vote percentages and outcomes would certainly extend the literature on shareholder activism by demonstrating that shareholder activism might not be solely predicted by financial interests.

Similar to other nonprofit organizations, labor unions find themselves in a relatively unique position as investors in and owners of firms, and their responsibilities are not necessarily limited to fiscal ones. This study is the first of its kind to show evidence that unions do, in fact, pursue nonmonetary motives through the vehicle of shareholder activism. Building on related research, this study adds an important layer to the picture of labor-union pension shareholder activism by looking at the firm as the unit of analysis. To better understand the firms that are (and are not) being targeted by union shareholder activists helps form a more complete picture not only of shareholder activism broadly but also of labor relations in the sphere of corporate governance. Most importantly, it helps to further add to the overall narrative of this study—that is, what do unions do *today*?

References

Bartley, T., and Child, C. (2014). Shaming the corporation: The social production of targets and the anti-sweatshop movement. *American Sociological Review, 79*(4), 653-679.

Benton, R.A. (2017). The Decline of Social Entrenchment: Social Network Cohesion and Board Responsiveness to Shareholder Activism. *Organization Science, 28*(2), 262-282.

Bizjak, J.M., and Marquette, C.J. (1998). Are shareholder proposals all bark and no bite? Evidence from shareholder resolutions to rescind poison pills. *Journal of Financial and Quantitative Analysis, 33*(4), 499-521.

Chava, S., Kumar, P., and Warga, A. (2010). Managerial agency and bond covenants. *The Review of Financial Studies, 23*(3), 1120-1148.

Cuñat, V., Gine, M., and Guadalupe, M. (2012). The vote is cast: The effect of corporate governance on shareholder value. *The journal of finance, 67*(5), 1943-1977.

Cutcher-Gershenfeld, J., Brooks, D., and Mulloy, M. (2015). *Inside the Ford-UAW transformation: Pivotal events in valuing work and delivering results.* Boston: MIT Press.

Daily, C.M., Dalton, D.R., and Cannella Jr, A.A. (2003). Corporate governance: Decades of dialogue and data. *Academy of management review, 28*(3), 371-382.

David, P., Bloom, M., and Hillman, A.J. (2007). Investor activism, managerial responsiveness, and corporate social performance. *Strategic Management Journal, 28*(1), 91-100.

Gillan, S., and Starks, L.T. (1998). A survey of shareholder activism: Motivation and empirical evidence.

Gillan, S. L., and Starks, L.T. (2000). Corporate governance proposals and shareholder activism: The role of institutional investors. *Journal of Financial Economics, 57*(2), 275-305.

Gillan, S. L., and Starks, L.T. (2007). The evolution of shareholder activism in the United States. *Journal of Applied Corporate Finance*, 19, 55-73.

Goranova, M., and Ryan, L.V. (2014). Shareholder activism: A multidisciplinary review. *Journal of Management, 40*(5), 1230-1268.

Hendricks, K.B., and Singhal, V.R. (2001). Firm characteristics, total quality management, and financial performance. *Journal of operations management, 19*(3), 269-285.

King, B.G., and Soule, S.A. (2007). Social movements as extra-institutional entrepreneurs: The effect of protests on stock price returns. *Administrative Science Quarterly, 52*(3), 413-442.

Kumar, P., and Ramchand, L. (2008). Takeovers, market monitoring, and international corporate governance. *The Rand Journal of Economics, 39*(3), 850-874.

Rezee, Z., Espahbodi, R., Espahbodi, P., and Espahbodi, H. (2012). Firm characteristics and stock price. *Abacus, 48*(4), 473-501.

Soule, S. A. (2009). *Contention and corporate social responsibility.* Cambridge University Press.

Vasi, I.B., and King, B.G. (2012). Social movements, risk perceptions, and economic outcomes: The effect of primary and secondary stakeholder activism on firms' perceived environmental risk and financial performance. *American Sociological Review, 77*(4), 573-596.

Webber, D. (2018). *The Rise of the Working-Class Shareholder: Labor's Last Best Weapon.* Harvard University Press.

Westphal, J.D., and Bednar, M.K. (2008). The pacification of institutional investors. *Administrative Science Quarterly, 53*(1), 29-72.

V. Building a Healthy Healthcare Workforce

Designing Work for Complex Care Management Programs: What Works for the Workers?

YAMINETTE DIAZ LINHART
CHRISTINE BISHOP
Brandeis University

Background

Healthcare systems pose unique employee challenges because care is often provided by cross-functional teams composed of professionals and lay healthcare workers from multiple disciplines. Healthcare workers such as community health workers (CHWs), patient navigators, and care coordinators are an emerging workforce integrated within health systems and accountable care organizations (ACOs) to improve health outcomes for patients with complex medical and social needs. Despite the challenges faced when working with complex patients, this population of healthcare workers, predominantly composed of women and underrepresented workers, has not been well-studied within the context of healthy work design and wellbeing.

The Centers for Disease Control and Prevention and the National Institute for Occupational Safety and Health (2002) highlighted the need for research to better understand how new work systems, such as ACOs, may affect workers, especially with respect to cross-functional teamwork, where the stress of working with high-risk and complex patients may lead to role conflicts, intensification of work, and changing skill-mix among workers (Sauter et al. 2002). Consistent with findings of health hazards in healthcare and social service workers, Holness et al. (2004) found that working with marginalized urban populations resulted in high workloads, increased stress, increased risk of violence and underreporting of occupational hazards. Given the high demands of the work, factors such as burnout, job dissatisfaction, low organizational and professional commitment, stress, and lack of social support are antecedents of turnover for social and human service workers (Mor Barak et al. 2001).

Although evidence points to greater occupational risks for these workers, there is little research available for this population of workers at the intersection of healthcare and social services. There have been three important national CHW studies from 1995–2010, and those studies have not focused on organizational factors in work design for safety and health of CHWs (Kash et al. 2007; Rosenthal et al. 2011). Additionally, CHWs serve as a unique population to study because they themselves reflect the characteristics of the patient population they are serving and are employed to address health disparities with this expertise. Employment of CHWs is expected to grow 11 percent by 2028, a higher-than-average projected growth for all occupations (US Bureau of Labor Statistics 2019b). Given the relatively new formal recognition of the CHW workforce as a distinct occupation by the US Department of Labor in 2010 and its projected growth, it is worth understanding the health and safety of CHWs within team-based care (US Bureau of Labor Statistics 2019a).

Research Questions

The overarching goal of this pilot case study is to describe organizational factors that impact the safety and wellbeing of healthcare workers employed to address the needs of high-risk complex patients in a complex care management (CCM) program of a Massachusetts Medicaid ACO and to generate hypotheses for further investigation:

1. What are the safety and health issues faced by healthcare workers in a complex care management program of a Massachusetts Medicaid ACO?
2. What are the organizational factors that contribute to workplace health and safety for healthcare workers in complex care management?

Methods and Data

This is a mixed-methods embedded case study of a medium-to-large size nonprofit ACO (50,000 to 150,000 covered lives). The ACO structure encompasses a large corporate parent organization that includes a managed care organization (MCO) and a joint venture with a smaller ACO in another part of the state. The ACO itself is composed of about 20 healthcare sites/organizations, including primary care clinics, community health centers, and hospitals in the eastern and western parts of Massachusetts.

Using purposive and convenience sampling, four types of data were collected and analyzed from participatory focus groups (n = 4 with 15 participants total across all focus groups), qualitative interviews (n = 32), field observations (n = 10 patient interactions) and survey responses (n = 60) to identify factors that contribute to safety and well-being for CHWs and other healthcare workers associated with CCM. Data were collected over a period of seven months from January to July 2019. Consistent with principles of participatory action research, we employed a co-design approach that included healthcare workers in the design of the research during the first two months of the grant period and after data collection was completed to validate findings in the final two months of the research (Bate and Robert 2006; Mulvale et al. 2016). Qualitative data was coded and analyzed in Atlas.ti software using a thematic analysis, and codes were reviewed on an ongoing basis by the authors (Ryan and Bernard 2003). Participatory synthesized member checking interviews and focus groups were conducted to validate themes at the end of the pilot study (Birt et al. 2016).

Findings

Findings from participatory focus groups revealed topic areas for research; these included challenges of team-based care, work stress and satisfaction, and organizational practices that support well-being of workers. As a result, the qualitative interview protocol included these topic areas along with questions to elicit description of day-to-day work and strategic decision-making process for CCM programs from administrators.

Qualitative interviews with CHWs, nurses, managers, and administrators revealed best practices of the CCM program as well as unique challenges posed by working in CCM. Table 1 summarizes qualitative themes. Workers described their work with patients as an important connection to their work. They also described, on the whole, feeling supported by their supervisors and organizations. These contribute to a sense of well-being and engaging in meaningful work.

> The most enjoyable thing that I find about this job is just continuing to work with the people that I've worked with over the past several years. It's like "oh my God, you're here now." Just seeing people make small strides, like I know a lot of times people want to see like huge improvement, and this is like no, this person can call their own transportation. That's huge, just doing something that they weren't doing before, able to instill dignity back into some people because some people don't feel like they deserve certain things and it's like yes, you do. (Interview I)

> There's definitely room for improvements but I feel very supported cause this is a great organization. I have nothing but good things to say. Yeah, I feel very, very supported. (Interview F)

Table 1. Qualitative Themes

Themes	Examples
Working Well	• Mission of the work (new day in healthcare, whole-person environment) • Supportive work structures (care planning, wellness groups, managers that understand)
Demands of Work Create Stress	• Time management issues due to nature of coordination and case load • Burdensome data and reporting requirements • Lack of control over space and resources
Integration and Coordination Challenges	• Sites don't always understand the role of CCM • Duplication of services across Medicaid care initiatives with community partners and other local community-based organizations
Opportunities and Challenges of Interprofessional Work	• Unclear roles and responsibilities between CCM workers cause stress • Team-based care is important for complexity of work
Commitment to the Work	• Satisfied and motivated workers despite work being stressful • Work variability is enjoyable • Relationships with patients anchor satisfaction and motivation for the work
Improvement Ideas	• Integrated and simplified reporting and data requirements • Organizational structures that facilitate integration and coordination to promote teamwork, coordination, resiliency, and sustainability • Measure benefits of relationship between CCM workers and patients for patient and organizational outcomes • Involve front-line workers in development of work practices (i.e., hiring, training) and programmatic decision making

Given the medical and social complexity of patients and the administrative changes as part of implementing an accountable model of care, the work demands themselves generate stress. Workers described time management, coordination across and within organizations, burdensome data requirements, and lack of control over work space as generating stress.

> It's a high caseload that I'm used to as a case manager, and it is challenging trying to put everything in place. This is set up as a short-term program for the average time of the year, but of course if there are ongoing needs you can carry someone a little longer, but it's challenging, it's challenging. It's a challenge. (Interview E)

> I think time management. That's definitely a stressor. We have to balance all these meetings just from our program, whether it's education or leadership

> meetings. We do like complex case conferencing, we're in a lot of meetings. And then, on top of that, we're balancing 37 patients that we're case managing intensely and then we're in care meetings and then meetings with other services, other organizations to set up care. And then we're also tracking these patients throughout all the levels of care. So, we check patients if they go inpatient at the emergency room in the hospital, discharged to the rehab facility, or back in the community. And on top of this, we're consistently documenting. And then, on top of that, we have to get people to enroll into the program, and then we have to discharge them and hand them off. So, I need to do all this at once. (Interview R)

Workers also described integration and coordination challenges within and across organizations; these included coordinating with multiple workers. Members, or patients, were also confused by the workers assigned to manage their care.

> So I have one patient who comes to mind because he's like my child. He had a social worker, he had VNA [visiting nurse association], he had PTOT [physical therapy and occupational therapy], all established when I started with him, and they basically knew nothing about the patient. They weren't doing anything for him. They were calling me. Once I started with him, and I helped him get his meds coordinated. I helped him with a pillbox and helped him understand how to fill them, helped him get connected to behavioral health therapy, helped him follow up with all of these specialists, figured out the right situation for that. Basically I was doing the job for them. (Interview G)

> And so what we see often is that we will say to the member, "We're going to work with your care team through the PCP's [primary care provider's] office to do all of these things for you," and they say, "This other person's already called me and told me they work with my PCP's office because they're trying to do the comprehensive assessment," and they are like, "If they work with my PCP's office, why don't they know the answers to these questions that they're asking me." Right? (Interview J)

Team-based care and interprofessional work is both challenging and important given the demands of the work. Appreciating the expertise of CHWs is an important opportunity for this workforce, which is often minimized due to their lack of credentialing in a highly credentialed environment.

> For me, my biggest challenge was knowing what a community health worker did. And now that I've been in this role and I've had a partner beside me who's a community health worker, they wear so many hats, and to me, they're actually really underestimated, and they deserve really high pay. I can't believe the amount of roles you play as a community health worker. I mean, you're a part nurse, you're a part therapist. I mean, you're a part care coordinator, you're a case manager. … In our data environment specifically, I really still have so much to learn about a lot of the social stuff. So, we really like to go together for our home visits. We try to do joint. All our patient care is joint, and it really helps each other to kind of remind each other, "Oh, was this done, was this done." She takes more of the social lead. (Interview R)

Finally, workers described feeling satisfied and committed to the work despite the stressors. Workers described the mission of the work as being key to their satisfaction, while also acknowledging the stress inherent in the work; in other words, workers described being both satisfied and stressed.

> I think it has to do with the mission of the work and the passion and the belief behind it. Like knowing this is something important and really believing it and kind of feeling like—it's worth it for me to dedicate my career to this kind of work. And it's going to be stressful and it's going to be under-resourced and a lot of public health things are, but the people around me inspire me and make it worth it, I think. (Interview D)

> I believe in the work we do, probably a depressing answer, but I just—I haven't found a way to like to do this work without it having like a significant negative impact on my own well-being, and it just hasn't been sustainable for me. But in terms of … how satisfied I am, I've recognized the things around this job that are good, and it's been part of … the puzzle for me. I'm thinking like, "Why am I … not happy in my work?" I get to do mission-based work. Like so many people in their professional lives, … they're working in a box factory. I get to … work at like the cutting edge of … Medicaid reform. Whenever I tell people what I do, they're like, "Oh, that's so cool." (Interview Q)

Results from the anonymous survey corroborated qualitative findings. Survey results included descriptive data from the National Institute for Occupational Safety and Health (NIOSH) Generic Job Stress questionnaire and Quality of Work Life questionnaire (CDC, "Organization of Work: Measurement Tools for Research and Practice," NIOSH 2017); Organizational Structures Assessment Tool (Jody Hoffer Gittell, forthcoming); Perceived Organizational Support Scale (Eisenberger et al. 1986); and the single-item burnout question adapted from Maslach's Burnout Inventory (Dolan et al. 2015). Some key findings from the anonymous online survey found that CCM workers experience high levels of stress, with 92 percent reporting their work as sometimes, often, or always stressful, while only 8 percent reported that their work was never stressful. Workers reported an average of 5.8 days of poor mental health in the previous 30 days (SD = 7.7). Using a 12-month lookback, 61 percent of workers reported having trouble going to sleep or staying asleep often or sometimes, while 39 percent reported rarely or never having trouble sleeping. Workers also reported high job satisfaction (86 percent very or somewhat satisfied and 14 percent not at all satisfied.) The single-item burnout question revealed that 38 percent of respondents endorsed "burnout symptoms" or "being burned out," while 49 percent reported being "occasionally under stress" and not having "as much energy" as once before, but "not feeling burned out," and 13 percent reported "enjoying work" and "no symptoms of burnout."

Overall, respondents reported feeling supported by their organization, with more than half of respondents consistently reporting feeling valued and supported by their organization on the Perceived Organizational Support Scale. Means and standard deviations using the Organizational Structures Assessment Tool, which measures relational coordination through existing organizational structures in eight human resource practices and four coordination mechanisms, were consistently rated as "sometimes," meaning that respondents rated roles as sometimes being hired and trained for coordination (as opposed to always or never) across all four professional roles (CHWs, nurses, physicians, and social workers). The only exception to this was "shared rewards (i.e., monetary recognitions for coordinating work. Sixty percent of respondents rated CHWs as not being hired for shared rewards (mean of 1.47, N = 40), and about 46 percent of respondents rated social workers as not being hired for shared rewards (mean of 1.61, n = 26). This finding is likely due to the lower wages of CHWs and social workers in hierarchical medical settings. Creating organizational structures that incentivize coordination through shared rewards may be key for these boundary-spanner roles.

When results were brought back to the workers for participatory member checking and validation of themes, workers described findings as "reflecting their experiences as workers." Workers also wanted to incorporate findings as part of their improvement processes. Unfortunately, by the time findings were reported back to the program leadership team for improvement ideas in March 2020, the COVID-19 pandemic shifted organizational priorities. However, worker well-being may be an imminent priority as healthcare systems continue to battle the pandemic.

Conclusion, Limitations, and Next Steps

What works for the workers? CCM workers voiced the need to support well-being through supportive work structures, including care planning meetings, supervision, and resiliency groups to debrief stresses of work. Work should be designed to reduce duplication of services, clarify roles and responsibilities, and help workers coordinate within and across different organizations. Organizations should include workers in both strategic and day-to-day decision-making processes to increase transparency and reduce stress of front-line CCM workers. Workers described the importance of employee-involved management practices and representation of worker voices throughout the organization.

> Some improvements, I'll just say one, and this is like high level. When making any decision about any type of program like this, always ask the people at the very, very bottom about how it is that they should go about to do anything because I find that higher-level administration and staff don't have a good picture of day-to-day how things are and about just people-to-people interaction. (Interview I)

> Our workforce is like 80 PERCENT women of color, and our leadership team is like basically all white people. So, like, there is representation issues, there's communication issues. ... how do we build a sense like throughout the layers of the organization that the organization is cohesive, because I think that's ... related to retention. (Interview Q)

Finally, this pilot study demonstrated THE feasibility and importance of research with CCM workers in Medicaid ACOs. CCM workers are emerging as a unique intersection of interprofessional workers tackling both medical and social health needs. Due to the demands of coordination and integration of the work, CCM workers devote substantial effort to system work-arounds and expect the work to be demanding and stressful. Workers also draw substantial satisfaction from the mission of the work, while also noting the emotional labor of caring for both patients and organizations they work with. Because of the nature of this research as a pilot case study, generalizability is limited; however, this research identified new questions to help understand, operationalize, and measure well-being for this workforce. Research on worker well-being is especially important given the demands placed on the healthcare workforce during the COVID-19 pandemic.

Acknowledgments

This research was sponsored by the Harvard Chan Education and Research Center Pilot Project Grant, No. T42 OH008416-14 from the National Institute for Occupational Safety and Health (NIOSH).

References

Bate, P., and Robert, G. (2006). Experience-based design: From redesigning the system around the patient to co-designing services with the patient. BMJ Quality and Safety, 15(5), 307–310. https://bit.ly/2HCQOBd

Birt, L., Scott, S., Cavers, D., Campbell, C., and Walter, F. (2016). Member Checking: A Tool to Enhance Trustworthiness or Merely a Nod to Validation? Qualitative Health Research, 26(13), 1802–1811. https://bit.ly/2HFlIsR

CDC. Organization of Work: Measurement Tools for Research and Practice. NIOSH. (2017, June 14). https://bit.ly/3mWJ7pB

Dolan, E.D., Mohr, D., Lempa, M., Joos, S., Fihn, S.D., Nelson, K.M., and Helfrich, C.D. (2015). Using a Single Item to Measure Burnout in Primary Care Staff: A Psychometric Evaluation. Journal of General Internal Medicine, 30(5), 582–587. https://bit.ly/3kTkEQ8

Eisenberger, R., Huntington, R., Hutchison, S., and Sowa, D. (1986). Perceived organizational support. Journal of Applied Psychology, 71(3), 500–507. https://bit.ly/30ukCqr

Holness, D.L., Somerville, S., Kosny, A., Gadeski, J., Mastandrea, J.J., and Sinclair, G.M. (2004). Workplace health and safety concerns in service organizations in the inner city. Journal of Urban Health: Bulletin of the New York Academy of Medicine, 81(3), 489–497. https://bit.ly/3mXC5kw

Jody Hoffer Gittell. (Forthcoming). Guidelines for Theory, Measurement and Analysis.

Kash, B.A., May, M.L., and Tai-Seale, M. (2007). Community health worker training and certification programs in the United States: Findings from a national survey. Health Policy (Amsterdam, Netherlands), 80(1), 32–42.

Mor Barak, M.E., Nissly, J.A., and Levin, A. (2001). Antecedents to Retention and Turnover among Child Welfare, Social Work, and Other Human Service Employees: What Can We Learn from Past Research? A Review and Metanalysis. Social Service Review, 75(4), 625–661. JSTOR. https://bit.ly/2S2ii50

Mulvale, A., Miatello, A., Hackett, C., and Mulvale, G. (2016). Applying experience-based co-design with vulnerable populations: Lessons from a systematic review of methods to involve patients, families and service providers in child and youth mental health service improvement. Patient Experience Journal, 3(1), 117–129. https://bit.ly/3i904d1

Rosenthal, E.L., Wiggins, N., Ingram, M., Mayfield-Johnson, S., and De Zapien, J.G. (2011). Community health workers then and now: An overview of national studies aimed at defining the field. The Journal of Ambulatory Care Management, 34(3), 247–259. https://bit.ly/3cB3t2O

Ryan, G.W., and Bernard, H.R. (2003). Techniques to Identify Themes. Field Methods, 15(1), 85–109. https://bit.ly/3cyjrLi

Sauter, S.L., Brightwell, W.S., Colligan, M.J., Hurrell, J.J., Katz, T.M., LeGrande, D.E., Lessin, N., Lippin, R.A., Lipscomb, J.A., Murphy, L.R., Peters, R.H., Keita, G.P., Robertson, S.R., Stellman, J.M., Swanson, N.G., and Tetrick, L.E. (2002). Knowledge Gaps and Research Directions (p. 43). Department of Health and Human Services. https://bit.ly/2S56yyU

U.S. Bureau of Labor Statistics. (2019a, May). Occupational Employment and Wages: 21-1094 Community Health Workers. https://bit.ly/337pZgM

U.S. Bureau of Labor Statistics. (2019b, May 9). Health Educators and Community Health Workers: Occupational Outlook Handbook. https://bit.ly/337t947

VI. Labor and Journalism in 2020

Intersectionality, Ideology, and the "Working Class": A Critical Appreciation of Christopher Martin's *No Longer Newsworthy*

MICHAEL HILLARD
University of Southern Maine

Introduction

What is the US "working class"? The connotations and use of this phrase have never been more loaded than in the age of Trump. With a keen eye on the 2020 national election, journalists, activists, political leaders, and concerned labor scholars have debated who is included under this label and how our understanding of it shapes electoral and policy outcomes, especially those impacting workers. Christopher Martin's recent *No Longer Newsworthy* is a timely, insightful intervention into this question. Martin is among recent labor scholars (e.g., Jefferson Cowie) who show how the death of the New Deal coalition and rise of neoliberalism in the 1980s produced a cultural shift in which the very concept of "the working class" disappeared from American political discourse—replaced by the politics of identity and progressive concerns about the "Reagan Democrat" and now the blue-collar, white base of the Trump coalition (Cowie 2010; Harvey 2006; Martin 2019). Martin offers a perceptive and carefully documented analysis of the decline of US labor journalism. Further, he investigates the media's now combustible representations of blue-collar America steeped in whiteness, while pointing the way toward a revival of a labor journalism that can serve as a catalyst for rebuilding sources of worker power and social justice.

This paper highlights Martin's two main thrusts. First, he documents the fall of the labor beat and institutional journalism's abandonment of the working class. Martin builds a powerful case that American journalism, especially newspapers and national television news, went from ably covering the working class and its institutions and issues to being cheerleaders of the 1 percent, aided and abetted by a new focus on the "upscale consumer" and the glamorous "entrepreneur." The leading sources of journalism allowed the growing disparities of wealth and income and the dramatic deterioration of economic life for workers to escape notice. In this, it has failed its essential role as a pillar of democracy—that is, the responsibility to "monitor power and offer a voice to the voiceless" (Kovach and Rosenstiel 2001: 130; cited by Martin 2019: 134).

Second, Martin interrogates how mainstream journalism in effect lost its muscle memory in understanding a working class that it no longer covered or spoke about. The telling example: The many failures of the media's coverage of "the biggest political story of our generation"—Trump's victory in the 2016 presidential election. Here, Martin first describes how Trump benefited from "two problematic ways in which the news has recently and historically framed its coverage of the working class" (Martin 2019: 134). First, coverage of the "working class" per se, and the Trump "base" in particular, is drained of any context; little is known or understood from media coverage about institutional forces shaping the economic and political life of workers. Instead, the press covers the "working class" narrowly through the lens of a political story devoid of expert context. Second, the press idealizes the working class as a narrow stereotype: white male blue-collar workers in "heartland" states. For Martin, this served Trump's interests neatly, as an example of how "divisive politicians ... exploit this image and divide working-class people on every other dimension: race, gender, sexual orientation, disability, and citizenship" (Martin 2019: 4). Martin then delivers a critique echoed by a number of excellent left-wing journalists, notably Sarah Jaffe and Mike Davis (Davis 2017; Jaffe 2020).

When Trump shocked the nation with his victory, leading journalists from the *New York Times*, *The New Yorker*, *Forbes*, *The Atlantic*, CBS News, and MSNBC (along with many other usual suspects) became obsessed with a "vexing question"—"Who are these people"? (Martin 2019: 1). Shortly after the election, leading journalists and some scholars set off on expeditions to "heartland states" that conjured up a connotation of "working class" that was reduced to a white male blue-collar worker in a Midwestern state who voted for Trump; problematic texts like Hillbilly Elegy became must reading for dumbstruck coastal journalists (Jaffe 2020; Martin 2019: 1; Vance 2016). Martin, Davis, and Jaffe all point out that this was an obsession with a very narrow demographic, legitimizing a sudden new language of class that idealized a relatively small slice of the electorate, those white, MAGA hat–wearing people at 2016 Trump rallies, a group that was, in fact, as Mike Davis succinctly describes, "Several hundred thousand white, blue-collar Obama voters, at most, voted for Trump's vision of fair trade and reindustrialization, not the millions usually invoked" (Davis 2017). Martin, especially through a close analysis of the "Carrier Corporation story" (more on this below), demonstrates just how much mainstream journalism missed in its obsession with unpacking this newly iconic Trump voter. He also shows what bad journalism this was as an explanation of the 2016 election.

We'll return in the second half of the paper to a reexamination of "the working class" and the 2016 election that Martin frames so well. Let's first look at the heart of his book.

Martin's *No Longer Newsworthy*: The Working Class Disappears From US Journalism and Political Discourse

Martin depicts the disappearance of the "labor beat" in newspapers and most national and local news organizations since 1970. Before 1970, newspapers large and small throughout the United States maintained a well-staffed labor beat. In that era, the press provided a widespread and informed coverage of workers, steeped in comfortable usage of the term "working class" and conducted by dedicated (I mean that in at least two senses) labor reporters who covered the challenges of work in mid-20th century capitalism. This journalism was appreciative of the role and necessity of unions in providing needed countervailing power against US employers with a long history of worker exploitation. Martin builds on his prior, equally excellent *Framed! Labor and the Corporate Media* to show the value this journalism once had and how a shift in the business model of newspapers roughly about 1970 led to the abandonment the labor beat (Martin 2003).

The media's abandonment by the press was spurred by national business currents that reshaped the news media's business model (Martin 2019: 61–67). Newspapers increasingly merged into national chains, run as publicly traded Wall Street based corporations, just when US business became focused more on a company's stock price rather than sales or stakeholder outcomes (Lazonick and O'Sullivan 2000). Competition from television winnowed out many newspapers with working-class readerships, and rural distribution was viewed as a cost center to be cut. As affluent workers moved into suburbs, and inequality per se became more pronounced, newspapers relied less on a broad, working-class readership for subscription revenues and increasingly relied on advertising revenues for higher-end retail chains, right at the time that domestic inequality began to disproportionately shift buying power to elites. Mainstream journalism thus moved to primarily "hail" the upscale consumer, evident in the new "economy" beats such as travel, personal finance, and fancy gastronomy. In the process, labor journalism eroded to a shell of its former self.

Why does this matter? In short, the new American journalism (with noted exceptions) that these changes ushered in gave no voice or visibility to workers, leaving our democracy being described in ways that ignored or supported unbridled corporate power. Going largely unexamined were new neoliberal policies that contributed to a profound concentration of wealth and poverty and an impoverished set of choices and conditions facing the US working class—the 60 to 65 percent of wage laborers who exercise little power in their jobs and lives (Zweig 2004: 4, 28–34; Zweig 2004: 4).

The loss of an effective labor beat means the failure of mainstream media to be a pillar of a democracy, the proverbial fourth estate. As noted, journalism scholars argue that journalism's fundamental goal is to "monitor power and offer a voice to the voiceless" (Kovach and Rosenstiel 2001: 130; cited by Martin 2019: 134). But this is generally not what news organizations do now, "especially when it involved exposing the uncomfortable issue of socioeconomic class" (Martin 2019: 134). Instead, journalism has been a passive and credulous witness to rise of a new, worker-unfriendly concentration of economic power in finance, tech, and retail sectors; in Martin's words: "chronicling—even advocating—the consolidation of corporate economic and political power, and sidelining the working class and labor" (2019: 134). Not surprisingly, this came as news corporations themselves were dismantling unions and retrenching, "while refocusing their priority audience from citizens to consumers" (Martin 2019: 134).

What could journalism have done to keep the powerful accountable? In Chapter Five, Martin uses his considerable skills as a communications scholar to document two interrelated trends of language—the abandonment by presidential candidates and presidents of a language of the working class, and its replacement with a neoliberal language of "consumers," "Wall Street," "investors," and "entrepreneurs"; journalists uncritically conveyed this new terminology behind which workers virtually disappeared. Lest we think it's just a matter of the neutral use of language, he demonstrates one of Marx's greatest insights, which is language that is ubiquitous and deployed by the institutional holders of power is itself a "material force," every bit as consequential as the use of state military and police power to suppress labor strikes or Black Lives Matter protests (Tucker 1978: 172–173). When labor journalism and unions were still a major force, a Republican president such as Eisenhower could celebrate that workers and their organizations were on an equal footing with their employers by saying, "Free collective bargaining without government interference is the cornerstone of the American philosophy of labor–management relations" (Martin, 2019: 137). Decades later, a southern Democratic president, Bill Clinton, adopted a different language, turning workers into objects to be trained in support of celebrated "entrepreneurs." Clinton's policy platform and speech abandoned any consideration of unions or a working class; rather, he celebrated entrepreneurship, arguing for an "expanding entrepreneurial economy of high-wage, high-skill jobs" (Martin, 2019: 141). Clinton continued a modified version of Reagan's hailing of so called working families: "Clinton's address included other tax cuts and credits for 'middle-income families,' 'every working family,' and 'every middle-income working family in this country" (Martin, 2019: 142). Martin concludes:

> From Eisenhower to Trump there has been a slow but dramatic tectonic shift in political discourse about labor unions and the working class. Presidents have lost the ability to speak of the essential role of labor unions in the US economy. … (T)hey've become true believers that tax cuts and regulations cuts are job creators; they've mostly become lost in a Mobius strip of crazy reason whereby one should not increase wages or benefits to make jobs sustainable because that would kill jobs; they decry the loss of community but kneel at the altar of individual when it comes to solutions. (Martin 2019: 146)

While political leaders abandoned the working class and unions, the discourse of the working class, labor, and unions was replaced by a new, upscale discourse celebrating the Horatio Algers and Rockefellers of our era (Martin did this by using Google Books' database of US English books from 1900 to 2008) (Martin 2019: 147–151). Again, the 1970s were the threshold. At that point, the term "citizen" was being replaced by "consumer"; the working class and middle class were bested by Wall Street; and by 2000, "entrepreneur" was more common than "working class", very much the result of media focus on "the 1 percent"—CEOs, finance wizards, and super-rich company founders in what one critic dubbed "wealth porn" (Poole 2000: 22; cited in Martin: 148).

Martin argues that mainstream journalism's abandonment of the working class, especially by newspapers and their post-1970 focus on upscale audiences, created an opening for conservative media to selectively fill. In recounting how Iowa voters supported Obama twice but moved sufficiently in Trump's direction to give him that state, Martin notes with irony the medium-sized Midwestern and Plains state newspapers such as the *Des Moines Register* that endorsed Clinton in 2016 (many breaking with their usual endorsement of Republican presidential candidates) as they recoiled in horror at Trump (Martin 2019: 154–156). But by 2016, even those papers spoke more to the educated urban/suburban voters in blue and purple states; white working-class voters in those states were instead turned into Fox News and Rush Limbaugh. Moreover, the media/consumer ecosystem created by the rise of the Fox News audience and its place among (some) working-class news consumers created a dynamic whereby mainstream journalists unwittingly embraced news frames created by Fox commentators' specious yellow journalistic advocacy. (More on this later in a discussion of Martin's Carrier Corporation case study.)

Indeed, awash in this congratulatory focus on celebrating the latest trends in American capitalism, mainstream journalism also slid into repeating the new conservative consensus in economic thought, a consensus that critics have shown to be derived not from empirical analysis but from theoretical deduction and which attacks countervailing power—such as minimum labor standards, the right to organize unions, progressive taxation, or environmental protections—as "interference" in "perfect markets" that leads to "suboptimal outcomes" (MacLean 2017; Cassidy 2009). A phalanx of traditional business lobbies (e.g., state and national Chambers of Commerce, the Business Roundtable), along with more extreme groups such as the complex of Koch political arms (e.g., Americans for Prosperity, the American Legislative Exchange Council), have used extreme conservative theory to buttress national and state campaigns to disarm or dismantle all forms of regulation and taxation of business (Lafer 2017; MacLean 2017; Phillips-Fein 2010).

In Chapter Six, "Job Killers in the News," Martin demonstrates how journalism became a propaganda arm of this movement by various sectors of American business. The focus here is the keystone of right-wing economic propaganda—the labeling of government countervailing power deployed in the regulation of labor standards or environmental protection as "job killers." According to Martin, this term became a "politically charged bludgeon" used to by "Republicans and business groups to attack government's legitimate regulatory role—protecting consumers, workers, public health, and the environment" (Martin 2019: 165). In the language of mass communication scholars, frequent media coverage of a topic is "agenda setting." Worse, uncritical adoption of a highly political "frame" normalizes spin as fact. When agenda setting and framing go unchallenged, they become propaganda, which "selects facts or invents them to serve the real purpose: persuasion and manipulation" (Martin 2019: 164). Propaganda and journalism are opposites; what journalists do if they do their job correctly is seek to verify facts and truth. This the media failed to do in the case of "job killers."

Martin, in collaboration with Peter Dreier of Occidental College, did a content study of the use of "job killer" over 1984–2011 by the *New York Times*, the AP, the *Washington Post*, and the *Wall Street Journal.* (Martin 2019; 167–178). Dreier and Martin found that journalists in 91.6 percent of some 381 stories using the term left it unchallenged—almost half of the stories arose in Obama's first three years in office. This journalism "of assertion" replaced "journalism of verification." The conclusion here is the problem got worse over time, and more partisan. Picking one vivid example, Martin reviews how the AP covered the 2010 failure of the Senate "cap and trade" carbon reduction bill. Because the AP feeds 1,700 newspapers and 5,000 radio and television outlets, and "without any journalistic attempt at verifying or presentation of competing claims," the frame that this bill was a "job killer" was repeated some 12,800 times. Thus, as deployers of this sloppy and unsubstantiated propaganda term, "news organizations primarily serve as transmission belts for those who oppose government regulation of business" (Martin 2019: 179).

In sum, by abandoning the labor beat, and in the economic sphere, the responsibility to "monitor power and offer a voice to the voiceless," journalism had come to abet corporate power over workers and delegitimize those institutions and policies that address workers' issues and struggles. As Martin concludes:

> As the mainstream media moved towards upscale audiences and reported on workplace lifestyles and personal finance from the 1970s onward, they left little room to consider the fate of American's working class in such an era of economic transformation. Meanwhile, the conservative media pushed a strong antiunion agenda in support of the Republican Party. And Democrats moved to the right as well, taking organized labor for granted. *A mainstream news media with an editorial focus on their entire community would have kept stories of the working class in the public sphere. But with no watchdog, politicians were left to wander away from a real commitment to the economic well-being of America's working class.* (Martin 2019: 147; emphasis added)

Covering Trump's Election, White Identity Journalism, and "The Working Class"

If the mainstream press abandoned labor journalism, what are we left with in its wake? A profoundly reactionary and deeply flawed analysis of Trump's unexpected 2016 election. With it came the sudden, alleged "discovery" by journalists that a disgruntled white, blue-collar, mostly male working class was the essential cause of Trump's election.

The "White Working Class" Was Not the Essential Cause of Trump's Election

In fact, there were more profound reasons for Clinton's loss. Martin demonstrates (as have others) that mainstream journalists are so lost at this point that they cannot seem to recognize that the predominant worker in the United States now is not in factories but rather in cubicles, caring, and retail labor (Draut 2018; Jaffe 2020). Moreover, workers are more typically low wage, lack benefits, or are in unreliable gig economy jobs—and in need of and striving for higher government labor standards and forms of countervailing power such as unions and worker centers that could realistically improve their fortunes (Draut 2018).

Attributing (or blaming) the shocking electoral outcome on this very specific demographic is a profound and lame failure of what academics would call an essentialism; that is, the media picked out one of a host of obvious factors that caused this electoral outcome while underplaying or ignoring others. At the top of a nonessentialist analysis was voter suppression. To pick one example, Wisconsin had a new voter ID law that reduced turnout by 200,000 votes in 2016 compared with the 2012 presidential election; Clinton lost Wisconsin in 2016 by a mere 22,748 (Jaffe 2020: 94). The Supreme Court's 2013 gutting of the Voting Rights Act in *Shelby v. Holder* resulted in 868 fewer polling places in predominantly Democratic precincts that were disproportionately people of color (Jaffe 2020: 94; Liptak 2013). Other factors accounting for

Trump turning Midwestern states who had voted for Obama in 2012: weak turnout among multiple Democratic constituencies including people of color and younger voters per se; Clinton's poor campaign strategy, particularly no visits to those "blue wall" states and her calling red state voters "deplorables"; Russian interference; and Attorney General Comey's "October Surprise" (Martin 2019: 3). Finally, Trump benefited from the typical windfall gained by an "outsider" candidate of the opposite party after a two-term Democratic presidency—but unusually and especially of the very large group of 18 million voters who saw both Trump and Clinton as "unfit," 69 percent voted for Trump (Jaffe 2020: 93). Finally, Mike Davis shows that the much discussed but overemphasized group of white, blue-collar workers and their families who shifted from Obama in 2012 to Trump in 2016 in swing states boiled down to some 20 or so counties—a tiny slice of the electoral surprise in 2016 and arguably a factor way down the list of previously cited developments explaining the 2016 presidential election (Davis 2017). Moreover, a significant group of workers in these "heartland" locales aren't voters because they're immigrants. As we've learned in the horrid story of Trump and meatpackers forcing these workers back into meatpacking plants only to come down with COVID-19 in large numbers, this group is largely workers of color (Greenhouse 2020; Lowrey 2020).

Accurately Covering the Working Class and the 2016 Election

As noted, the correct formulation of what the current US working class is boils down to recognizing that only a small slice fits the dominant trope adopted by Trump and reified by journalists before and after Trump's victory. About one in ten workers are in factory jobs; roughly 40 percent are in retail and service (heavily low-wage healthcare) jobs (Durant 2018; Zweig 2000: 29). The white/workers of color and male/female (as counted by official statistics) split is about 50/50, particularly for workers under 40 years old (Durant 2018). Median earnings for workers in these jobs range from $20,000 to $30,000, few have benefits, they are concentrated in modes of employment that labor laws poorly protect if at all, and they offer little or no upward mobility (Durant 2018).

Crucially, improvements in these workers' lives do not come from the Great White Savior (i.e., Trump) but from organizing—activism like the Fight for $15, which led to large increases in state minimum wages across the United States, and the notable progress of worker center organizations such as the Coalition of Immokolee Workers as well as progressive feminist unions such as the National Nurses Association and the Harvard Clerical and Technical Workers Union (Estabrook 2018; Hoerr 1997; Rolf 2016). Martin's contribution, as is that by standout (if rare) labor journalists including Sarah Jaffe, Stephen Greenhouse, and Hamilton Nolan, is to identify frames that produce a good "journalism of the working class" that is accurate about its diversity and appreciative of what forces cause its suffering or progress. Martin thus points up the underlying political economy that can help tee up good investigative reporting—to see the that corporations not progressive policy activists are job killers, and that it is the agency of workers themselves with supportive government policies that can uplift the vast, precarious working class that Trump and neoliberal Democratic predecessors such as Bill Clinton have done little to assist. And this work has to address the structural inequities of race and gender, requiring a broader conversation of policy initiatives that until recently has been treated as the exotic and not all that American musings of socialist candidates.

And what to make of the post 2016 election "journalistic safaris" that sought to explain the shocking emergence of Trump? Martin compellingly argues that an effective labor beat would not have fallen into this journalistic trap. Recall that Martin and others point to the news media's unfortunate seduction by the narrative of "those people"—white, male, blue-collar workers—who suddenly switched to Trump and delivered him victory. As Sarah Jaffe argues, white Trump voters assumed to be working class were largely not (Jaffe 2020: 91–92). One simple error came from assuming that a white "heartland" voter without a college degree was "working class"; about "two thirds of Trump's supporters made above the household median income" of $57,600. Most telling is that, according to two *Washington Post* reporters, about 60 percent of white Trump voters without a college degree were in the top half of American households by income (Carnes and Lupo 2017). And as noted, an important part of the American blue-collar workforce is immigrants, many of whom are not citizens and can't vote.

And finally, according to Jaffe, there is a profound danger in holding up the white, male, blue-collar worker as "the" quintessential worker or member of the working class.[1] She calls this the "whitewashing of the working class." Journalists and the public have turned "white working class" into an identity category, bereft of understanding class based on "relationship(s) of power, one's relationship to the workplace and the means of production." If working class and white go together, "the corollary is that nonwhite people are not working"—the nonwhite are poor "takers" who live off of welfare and form an "underclass." "In this essentialized notion of the white working class, class is just something you are, as immutable as the color of your skin." Thus, white people in between the coasts or those with "blue-collar ancestry, are considered working class forever," even if they rise to become business owners or professionals. Furthermore, there is an assumption of a fixed and shared political–cultural makeup: "Conservatism on issues of gender and sexuality, a deep-seated antipathy toward all nonwhite people, and a distaste for environmentalism, the arts, or anything else popular with latte-sipping coastal liberals." For neoliberal Democrats (especially the Clintons), this adds up to the working class that can be categorized as "deplorables." Jaffe argues convincingly that the core of Clinton's poor or overall lack of campaigning in swing states was just this dismissal—one that contributed to her electoral loss.

Indeed, Clinton's failure was to reify a whole category of white reactionaries as "working class" and then comfortably abandon them in pursuit of the sensitive and socially liberal white, middle-class suburban voter who would be turned off by Trump's racism. Her disinterest in white voters outside of upscale cities and suburbs, and her taking the union vote for granted—and notably her deep underperformance among union voters in those swing states that showed a profound failure of strategy—speaks to the neoliberal identity of the candidate herself. In short, a whiteness reading of the category "working class" stands in the way of actually recognizing the working class for what it is—a diverse demography of largely nonmanufacturing workers whose status of working class is derived not from their love of NASCAR and dislike of LGBTQ communities but who are, in fact, those who sell their time for a wage, make a profit for their employer, and lack power in their workplaces.

So, Martin along with these other writers, demonstrates the folly and empirical weakness of whitewashing the working class in service of explaining Trump's surprise victory. Martin's specific contribution to this critique is his diagnosis of how contemporary journalists, lacking the conceptual tools to understand structural realities facing workers and uncritically susceptible to picking up right-wing frames, messed up in its lack of nuance and confusion (in ways laid out so well by Jaffe) of the intersection of class and Trump's election. Martin's analysis on this point is sharpest in his careful deconstruction of the biggest "worker story" associated with Trump—his role as an alleged savior of white, blue-collar jobs at the Carrier furnace plant in Indianapolis.

Carrier Corporation

In the period when Trump was president-elect, no story loomed larger than his alleged rescue of blue-collar jobs at Carrier in Indianapolis, a furnace factory and subsidiary of conglomerate giant United Technologies Corporation (UTC)[2]. UTC announced in February 2016 that it was shutting the operation within the year and moving production and 1,400 jobs to a Mexico facility. This story became a central piece of Trump's campaign meant to demonstrate his commitment (of sorts) to rescue the American factory worker from capital flight and globalization. Trump appropriated this devastating development, and his grandiose plans to interrupt it and save these workers, to build his "cred" as the man who would end this kind of "carnage." This story became, in Martin's words, "the central economic parable of his campaign" (Martin 2019: 23). Martin uses the story as an object lesson in demonstrating journalists' inability to set context or dig behind the stereotype of the working-class that the media was quickly building.

First the story. A worker videoed the plant manager's lame and hurtful announcement of the factory's impending closing and move to Mexico. The video went viral and was seen by millions, and it was reinforced by paleo-conservative Steve Bannon (then of Breitbart) and Murdoch's *New York Post.* It caught Trump's attention, and he built his planned rescue into his stump speech and debate talking points. He painted a scenario where he would, in strongman fashion, use the threat of a 35 percent tax on "air conditioners" (Trump didn't bother to get the actual product correct) to get the CEO to reverse his decision; his trope included the image of the weak CEO bending to Trump's manly will. But once elected, Trump forgot all about it.

Indeed, many Carrier workers and also workers at Rexnord, a nearby ball-bearings plant that was also planning at the time to move hundreds of jobs to Mexico, had voted for Trump. But, many did not—in fact, a large group were Sanders supporters. Trump's interest—he lost track of his commitments between spring 2016 and his election—was revived by an NBC news reporter whose November 14, 2016, report aired on the NBC Nightly News. Union official T.J. Bray—a white man—joined two other white workers in saying that they were disappointed to not hear from Trump on his campaign boast to save their day. When Bray mentioned (as he did in an earlier national interview) that he voted for Trump and "would vote for him again" if Trump saved their jobs, Trump's ego was sufficiently stroked to push him to swing into action. His campaign then used the convenience of Vice President-Elect Mike Pence's status as governor of Indiana to negotiate what in effect was a massive $7 million bribe of local financing and tax breaks to secure what turned out to be a portion of Carrier's jobs (and none at nearby Rexnord).

What happened next followed a template that has allowed right-wing tropes to become mainstream news coverage. The narrative form a story takes is its frame, and repetition of a frame by the media becomes "agenda setting" (Martin 2019: 164). This gets done by right-wing media, in this case Breitbart and Fox News. While mainstream media's journalism's obligation is to verify claims and question whether a frame is accurate or is spin, the failure to do so and further repetition of a frame turns journalistic story making into propaganda (Martin 2019: 164). That's in fact what happened.

Fox News and the *New York Post* set the frame—white male, blue- collar, union workers grateful to Trump the savior. It began with a Megyn Kelly interview on November 30, 2016, the day before Trump's visit to Carrier to announce the deal to save workers' jobs The focus of the interview: Kelly prompting the two men (Bray and another white male worker, Paul Roell) to say how grateful they were to Trump and to indicate they were supporters. The same night, Sean Hannity went one better and prompted another white male worker, Robin Maynard, to agree to a hagiographic screed about Trump's historic leadership abilities. While not quite rising to the level of Fox's sycophancy, national stories on NBC, CBS, and ABC news bought and repeated Fox's frame.

Like with Trump's electoral victory, mainstream journalists' coverage of Trump's deal making to keep some Carrier jobs in Indiana reified the quintessential working-class voter as white, male, blue-collar, unionized, and having moved firmly into Trump's base. And yet Carrier, like the American working class as a whole, is not majority white male. Indeed, Carrier's workforce in 2016 was 60 percent black and 40 percent female (Martin 2019: 31). The United Steelworkers' Local 1999, which represented these workers, was governed by a board of eight black and three white officials.[3]

Martin deconstructs the errors and flaws in Trump's speech to Carrier employees in his much-covered visit to the Indianapolis factory—errors and flaws that journalists completely missed. The flaws:

- Trump had completely forgotten about the Carrier plant until the national media reminded him of his repeated promises to rescue it.
- Trump had claimed he we would get the concession of keeping the plant in the United States from the UTC CEO by strong-arming UTC with the threat of a 35 percent tariff on products shipped to the United States from the new Mexico plant, when, in fact, Pence arranged for a $7 million taxpayer giveaway to bribe Carrier to stay in Indianapolis.
- Trump continued to mistake the plant (as did some Fox anchors) as an air-conditioner plant; it actually made furnaces.
- Trump couldn't remember T.J. Bray's name or what he looked like, though he called to him as a "gentleman, worker, great guy, handsome guy."
- USW Local 1999 got no mention in Trump's speech, even though it had been extremely active over the prior nine months in protesting and publicizing their efforts to save their plant, thus being the origin of national coverage, while Trump himself only negotiated with UTC CEO Gregory Hayes.
- Trump ignored that a total of 2,400 of factory jobs in the area (including another nearby UTC factory and Rexnord Bearings) were initially planned to be shipped to Mexico in announcements that year, focusing only on Carrier in Indianapolis.
- Trump dissembled in claiming that his negotiated deal would save all 1,100 Carrier jobs, when the number was actually 730 (disclosed after Trump's visit) (Martin 2019: 39).

By whiffing on any substantive verification of Trump's bombast, the media got to "tell Trump's less complicated story, no matter the facts: that the (white, male) working class put Trump in the White House, and now Trump was putting them back to work, just as he promised" (Martin 2019: 39). And when Local 1999 President Chuck Jones called Trump out in December for telling the workers on December 1 a falsehood about the number of jobs Trump's deal actually saved, Trump hate-tweeted Jones, assuring that his broad base knew that the union and its leadership were "terrible" (Martin 2019: 40). Meanwhile, national polling on Trump shot up on his alleged commitment to look out for the worker and save jobs from moving overseas.

The lesson here? Foremost, Local 1999's actions and role in the matter deserved factual attention that it didn't receive—a direct outcome of the absence of a labor beat in Indiana or nationally. The local had granted UTC major concessions; it had rallied at both Carrier and at the Indiana statehouse. Local 1999 was deft at publicizing the matter in social media. Its activism kept the story in the news, overcoming Trump's actual indifference. A large portion of Carrier's workers were in fact Bernie Sanders supporters, and Local 1999 had endorsed Sanders in the primary. Moreover, as is often the case, at the time UTC announced its plan to shutter the plant, it was highly profitable—just not as profitable as a Mexican plant with $3 per hour wages. And of course, its CEO was at the same time cashing in on massive earnings ($13.4 million in 2016), and UTC had the resources to issue the previous CEO, let go for corrupt practices, a $200 million "parachute." And, as with the 2016 election story, Martin points out that the national media covered Carrier as a political story and not a labor story, but rather one based on a trope that had been first served up by a fawning Breitbart and Fox evening anchors and "largely … picked up by the other national news media" (Martin 2019: 43). And Martin continues:

> The news media's general focus on white male Carrier workers as subjects denied the fact that the Carrier workforce was far more diverse in terms of gender, race, and politics than the role they were given in the story (as white,

> male, ardent Trump supporters). In the Trumpian economic populism, the white male Carrier workers the news media featured, to the exclusion of others, suggested which Americans have jobs worth saving. (Martin 2019: 43)

As Martin observes, what also screams out about this story, and what we've learned in the past four years, is that Trump's episodic deal making is not a policy, notable in the fact that the deal saved only about on third of the 2,400 factory jobs that were to be moved from the area to Mexico and that after this one deal, one is hard pressed to recall when he's repeated this one-time act of showmanship. In turn, Martin catalogs five substantive US House and Senate proposals (still languishing because of McConnell's disinterest in putting any constraints on the actions of US corporations) that have gotten no coverage in the media. And to drive home a central thread in *No Longer Newsworthy*, news reporting views "displacement by outsourcing as common-sense economic inevitability. The stories framed the usually white workers as pitiable characters, casualties of a global economy over which they have no control" (Martin 2019: 45).

Conclusion: Martin's Timely Agenda for Reviving Labor Journalism

Fortunately, Martin provides us with a road map to a better 21st-century journalism that recognizes the working class fully and correctly, avoids the whiteness trap outlined here, and ably describes and explains for a mass citizenry the conditions and issues that LERA scholars so ably analyze. Following Martin's excellent intervention, we can now see clearly that the United States needs more trained and dedicated (again in two senses) journalists who unearth abuse and exploitation, identify the ways in which untrammeled corporate power hurts worker, put a moral price on upscale consumerism by unearthing the social and economic conditions under which goods are produced and services are provided, and point to where and how a restoration of countervailing power can actually improve working-class life.

Also, at a meta level, what undergirds Martin's analysis is showing how profoundly what might be best described as how neoliberal ideology or the language of orthodox economics has seeped in and become the wellspring from which mainstream journalists operate. Similarly, the wellspring for Martin's analysis is a political economy (heterodox, and consistent with the institutional labor economics tradition) that recognizes the existences of classes, understood by presenting workers as what they are—waged workers, who in the absence of organization have had to submit to available job opportunities that come with too much danger and stress (even more so during the COVID-19 pandemic) and who receive too little compensation or security for their often arduous efforts.

Christopher Martin's N*o Longer Newsworthy* is a critical accomplishment that deserves to be widely read and discussed by labor/employment relations scholars and students. It is carefully researched, profoundly insightful, and beautifully argued. It should be taught in many graduate and undergraduate classes and should be the first present we give to any and all of our friends who work as journalists.

Endnotes

[1]All quotes in this paragraph come from Jaffe (2020: 95–97).

[2]This section summarizes Martin's chapter on Carrier (Martin 2019:47–68).

[3]Martin adds a detailed caveat on a number of national and local reporters who did represent the demography of Carrier workers accurately. This is a thread that runs throughout Martin's book. His repeatedly makes the point that while good journalists do some good stories, it doesn't add up to a consistency that would come from a dedicated labor beat (Martin 2019: 32–33).

References

Carnes, Nicholas and Lupu, Noam. 2017. "It's Time to Bust the My: Most Trump Voters Were Not Working Class. The Washington Post. June 5.

Cassidy, John. 2009. How Markets Fail: The Logic of Economic Calamities. New York: Farrar, Straus and Giroux.

Cowie, Jefferson. 2012. Stayin' Alive: the 1970s and the Last Days of the Working Class. New York: The New Press.

Davis, Mike. 2017. "The Great God Trump and the White Working Class" Jacobin, Feb 7.

Draut, Tamara. 2018 "Understanding the Working Class." Demos April 16, 2018

Estabrook, Barry. 2018. Tomatoland: From Harvest of Shame to Harvest of Hope. Kansas City: MO: Andrews McMeel.

Greenhouse, Stephen. 2020. "Trump is Marching Meatpacking Workers to Their Deaths." The Guardian, April 30.

Harvey, David. 2005. A brief history of neoliberalism. New York: Oxford University Press.

Hoerr, John. 1997. We Can't Eat Prestige: The Women Who Organized Harvard. Philadelphia, PA: Temple University Press.

Jaffe, Sarah. 2020. "Whose Class Is It Anyway? The 'White Working Class' and the Myth of Trump." In Jasmine Kerrissey et al., ed., Labor in the Time of Trump. Ithaca NY: ILR Press, pp. 87–105.

Kovach, Bill, and Rosenstiel, Tom. 2001. The Elements of Journalism. New York: Three Rivers.

Lafer, Gordon. 2017. The One Percent Solution: How Corporations Are Remaking America One State at a Time. Ithaca, NY: ILR Press.

Lazonick, William, and O'Sullivan," Mary. 2000. Maximizing Shareholder Value: A New Ideology for Corporate Governance," Economy and Society, Vol. 29, no. 2, pp. 13¬¬–35.

Liptak, Adam. 2013. "Supreme Court Invalidates Key Provision of the Voting Rights Act." The New York Times, June 25.

Lowrey, Annie. 2020. "Don't Blame Econ 101 for the Plight of Essential Workers. They've been systematically devalued for years. But they don't have to be." The Atlantic, May 13.

MacLean, Nancy. 2017. Democracy in Chains: The Deep History of the Radical Right's Stealth Plan for America. New York: Penguin.

Martin, Christopher C. 2003. Framed! Labor and the Corporate Media. Ithaca, NY: ILR Press.

Martin, Christopher C. 2019. No Longer Newsworthy: How the Mainstream Media Abandoned the Working Class. Ithaca, NY: ILR Press.

Phillips–Fein, Kim. 2010. Invisible Hands: The Businessmen's Crusade Against the New Deal. New York: W.W. Norton.

Poole, Gary Andrew. 2000. "'Wealth Porn' and Beyond." Columbia Journalism Review Vol. 39, (November/December), pp. 22–23.

Rolf, David. 2016 Fight for Fifteen: The Right Wage for a Working America. New York: New Press.

Tucker, Robert C. 1978, ed. The Marx-Engels Reader. 3rd edition. ¬New York: Norton.

Vance, J.D. 2016 Hillbilly Elegy: A Memoir of a Family and Culture in Crisis. New York: Harper.

Zweig, Michael. 2000. The Working Class Majority: America's Best Kept Secret. Ithaca, NY: ILR Press.

Zweig, Michael. 2004. "Introduction: The Challenges of Working Class Studies." In Zweig, ed., What's Class Got to Do With It? Ithaca, NY: Cornell University Press.

VII. Building Multi-Stakeholder Collaborations in Education

Educator Coproduction with Parents: The Case of Elementary Education of Children of Immigrants with Special Needs

CADY LANDA
Brandeis University

Parent engagement in their children's education is widely acknowledged to impact students in positive ways, but there is a wide variety of methods by which parents are and can be engaged. Schools have inherited bureaucratic structures that typically result in hierarchical relationships between educators and parents that position parents to play a subordinate, peripheral role in their children's education. Application of a relational coproduction model in school would feature structures supporting educators in having a more reciprocal role relationship with the parents of each of their students. School structures often do not provide supports to educators to engage with their students' parents to this extent, however. Even more rarely do schools make a significant investment to ensure that parents have the resources they need to partner with educators in the delivery of educational services to their children.

Because the quality of educational services and student outcomes are influenced by the quality and level of parent participation and because there is heterogeneity among parents with respect to the resources they have to participate in their children's education, failure of school structures to remedy significant differences in the resources parents have to participate can create inequities in service delivery to students that replicate or even widen, for the students, the disparities that existed in their parents' resources.

Because structural factors leave so many immigrant parents without many of the resources needed for effective coproduction with educators on behalf of their children, the children of immigrants is one group of students who are at risk of experiencing inferior educational services. This paper presents the methods and results of a mixed methods study that looks at the access of children of immigrants in elementary school to special education entitlements and other school services addressing needs of students having difficulty. A quantitative analysis of state administrative data asks whether children of immigrants are less likely than children of US-born parents to receive special education and whether, among students who do receive special education, children of immigrants are more likely than children of US-born parents to be served in substantially separate settings and less likely than children of US-born parents to be served in inclusive settings. A companion case study of an elementary school in the same state explores how and why the outcomes observed in the quantitative analysis occur. Research questions for the case study ask the following:

- How do school organizational structures and public policies affect the ability of the school staff to meet special needs of children of immigrants with low income and work with their parents in the process of doing so?
- What are the patterns of parent experience navigating the school system on behalf of their children having difficulty in school? What are the influences that shape those experiences in particular ways?

Exploring the educational equity of students of immigrants is important because of their large number and share of the US student population and the public policies that marginalize or fail to integrate so many of their parents. The number and proportion of children in the United States with at least one immigrant parent are significant and have rapidly grown. In 2017, a quarter of all children in the United States had at least one immigrant parent. Between 1990 and 20017, the number of children with an immigrant parent more than doubled, from 8.3 million in 1990 to 19.6 million in 2017 (Child Trends 2018; Fortuny, Hernandez, and Chaudry 2010). In addition to the growing number of immigrants' children in school, the increased dispersion of immigrants away from traditional gateway areas has meant that an increasing number of education jurisdictions have responsibility for educating their children. This study focuses on children of immigrants who are struggling in elementary school. In so doing, it can include public policy and practice issues that affect the ability of schools to meet the needs of children of immigrants who are likely to be at greatest risk of adverse school experience and for whom parental involvement and strong educator–parent relationships can be particularly critical. Adverse experience in elementary school challenges children's developing concepts of self and can have detrimental impact on later academic and career trajectories (Coll and Marks 2009; Eccles 1999; Huston and Ripke 2006; Welsh, Parke, Widaman, and O'Neil 2001).

Together, the quantitative analysis and case study add to the literature in important ways. The quantitative analysis adds to the few studies that have examined comparative rates of special education participation among students with and without immigrant parents. These previous studies provide some evidence that children of immigrants at the elementary level are less likely to receive special education than children of US-born parents (Child Trends 2015; Conger and Grigorenko 2010; Hibel and Jasper 2012). No previous study has examined differences in rates of inclusive and substantially separate placements for children with immigrant and US-born parents.

In addition to adding to our understanding the mechanisms that contribute to the lower rates of participation of young children of immigrants found in quantitative studies, the case study contributes to the literature by focusing on malleable public policy and organizational factors that affect the ways in which educators and immigrant parents can work together to address the educational and developmental needs of young, at-risk students who are children of immigrants. Use of ecological, organizational, and immigrant adaptation theories adds to the literature by treating school organizational structure, and the ways in which it interacts with public policies affecting schools and immigrants, as an important influence on the developmental context of children of immigrants. The case study is unusual in including both students who do and do not receive special education in a single sample of interest. Doing so provides an opportunity to examine how the Child Find mandate of the Individuals with Disabilities Education Act (IDEA), requiring state and local education agencies to identify and offer services to all eligible students, is implemented in the case of students with immigrant parents. The case study also adds to a small body of literature on parenting as an immigrant and does so with a focus on parenting for young children experiencing difficulty at school.

Theoretical Frameworks

The quantitative analysis and case study share a constellation of theoretical frameworks that shaped hypotheses for the quantitative analysis and data collection and analyses in the case study. I used Bronfenbrenner's (1979) ecological model of human development as the overarching theoretical framework to examine the developmental context of study students. In the ecological model, the development of the individual occurs through interactions between individuals and their immediate environment within four nested systems. From most proximal to distal, these systems are as follows:

- Microsystem: the individual's immediate settings, which include, in this study, the parents and the child's school
- Mesosystem: the interrelations among the individual's immediate settings, which in this study are the interactions between educators and parents;
- Exosystem: the formal and informal social structures not directly experienced by the individual that affect features of their immediate settings—in this study, the school district, parent's social network, and parent's employer.
- Macrosystem: the socioeconomic, historical, and cultural contexts and public policies that influence all other levels.

The study uses relational coordination and relational bureaucracy theories to examine relationships of staff within the school (microsystem) and with students' parents (mesosystem) and how both are shaped by the school's organizational structure. Relational coordination theory holds that effective coordination is carried out through relationships of shared goals, shared knowledge, and mutual respect that are reinforced by communication that is frequent, timely, accurate, and problem solving (Gittell 2006). It is hypothesized that relational coordination results in beneficial outcomes for multiple stakeholders when it is strong and problematic outcomes when it is weak, especially as work becomes more interdependent, uncertain, and time constrained (Gittell 2003).

Relational bureaucracy, a combination of bureaucratic and relational organizational forms, extends relational coordination theory to propose that structures can be designed to encourage reciprocity across work roles based on shared goals, shared knowledge, and mutual respect among staff (relational coordination), between staff and leaders (relational leadership), and between staff and their clients (relational coproduction). The theory hypothesizes that these three kinds of reciprocal interactions foster an attentiveness to the situation and to one another and allow for an integration of perspectives that can produce caring, timely, and knowledgeable responses to the particular individuals served by the organization (Gittell and Douglass 2012).

The study uses Alba and Nee's (2003) immigrant adaptation theory to understand the experience of the immigrant parent (microsystem). This theory portrays individuals' adaptation to the new home as the unintended and contingent result of individuals' purposive actions, informed by their networks, knowledge and experience, and institutional incentives and constraints.

Quantitative Analysis Methods

Data

The Massachusetts Department of Elementary and Secondary Education (DESE) provided the October 2013 file of its Student Information Management System (SIMS) for this study. SIMS is the state's student-level data collection system administered by DESE to meet state and federal reporting requirements and inform policy and program decisions. I chose the 2013 file because of changes in the way SIMS measures student low-income status after 2013, which made more recent files more problematic for this research.

Samples

I drew two samples from the October 2013 SIMS: one for analysis of special education participation and the other for analysis of special education placement. The sample for analysis of participation includes all K–5 students who were coded as enrolled by their reporting district on October 1, 2013 (N = 425,538). The sample for analysis of placement (N = 57,075) includes only those students in the first sample who received special education on October 1 and were at least six years old.

Measures

Dependent Variables

The dependent variable for analysis of special education participation is a dichotomous variable coded "1" if the student received special education on October 1, 2013. There are two dependent variables for the analysis of placement:

- "Included," coded "1" for students receiving special education in their school's general classroom for at least 80 percent of the time
- "Substantially separate," coded "1" for students receiving special education in their public school but outside of the general classroom more than 60 percent of the time

Independent Variable of Interest

Because SIMS does not identify students who have an immigrant parent, this study uses a proxy and considers only those students noted in SIMS to have a first language other than English or Spanish to be a child of an immigrant. The expectation is that use of this proxy will err in the direction of more children of immigrants being considered children of US-born parents rather than the converse, making it more difficult to disprove the null hypothesis.

Control Variables

Covariates include student-level variables that are present in SIMS that are likely or have been shown to be correlated with special education participation and placement. These variables are gender, low income, race/ethnicity, limited English proficiency, Section 504 accommodation, and grade level. I use primary disability as an additional covariate in analyses of placement.

Statistical Analysis

This study uses Stata (Version 14.0) to estimate mixed effects logistic regression equations with random effects for school districts for each of the dependent variables. The analysis includes random effects for school districts in addition to control variables because state policy tolerates a degree of variation among school districts in their administration of special education and Heher et al.'s studies of special education in MA (2012; 2014) found significant differences at the district level in rates of student participation in special education, attribution of disability type, and inclusiveness of student placement. I report the results of the fixed effects as odd ratios.

I ran regressions with the parent nativity variable (Model 1) and without the parent nativity variable (Model 2) to clarify the impact of including a variable for parent nativity on relationships between the dependent and the other control variables, typically used to measure disparities in education administration.

Results of the Quantitative Analysis

The quantitative analysis strongly suggests that having an immigrant parent has a significant independent and negative impact on young children's access to special education and a significant negative impact on access to inclusive placement for those young children who do receive special education.

Descriptive Statistics

Table 1 displays the distributions of dependent and control variables for the populations examined in this study: those with and without the proxy for having an immigrant parent. Prior to controlling for the other correlates, students identified as children of immigrants had a lower rate of participation in special education (10.2 vs. 16 percent) and among those receiving special education, a lower rate of full inclusion (65 vs. 69.4 percent) and a higher rate of substantially separate placement (17.9 vs. 13.7 percent) than students identified as having US-born parents.

Table 1. Variable Proportions by Proxy for Having an Immigrant Parent

Variable	**Students with Immigrant Parent Proxy N = 40,855**	**Students with US-Born Parent Proxy N = 384,683**	**P value**
Male %	52.0	51.3	<.01
Low Income %	57.2	38.6	<.001
Race/Ethnicity			
African American/Black %	20.2	7.1	
Asian %	38.3	3.1	<.001
Hispanic/Latino %	8.9	19.8	<.001
Native American %	0.5	0.2	<.001
White %	30.0	60.0	<.001
Native Hawaiian/Pacific Islander %	0.2	0.1	<.001
Multi Race, Non-Hispanic %	1.9	3.6	<.001
Limited English Proficiency %	53.9	6.5	<.001
Section 504 Plan %	1.0	2.8	<.001
Receiving Special Education %	10.2	16.0	<.001
Fully Included %	65.1	69.4	<.001
Substantially Separate %	17.9	13.7	<.001
Primary Disability			
Intellectual %	4.1	2.9	<.001
Hard of Hearing/Deaf %	1.6	0.7	<.001
Communication %	31.2	23.4	<.001
Vision Impairment/Blind %	0.6	0.4	<.05
Emotional %	3.2	5.5	<.001
Physical %	1.4	1.4	.664
Health %	4.4	9.0	<.001
Specific Learning Disabilities %	15.9	21.0	<.001
Deaf and Blind %	0.4	0.1	<.001
Multiple Disabilities %	1.7	2.1	.116
Autism %	11.0	10.4	.327
Neurological %	4.8	4.2	.094
Developmental Delay %	19.8	18.8	.131

Note: Students with the proxy for having an immigrant parent have a first language other than English or Spanish. Students with the proxy for having a US-born parent have English or Spanish as a first language.

Special Education Participation

Table 2 provides the results of the multivariate analysis of special education participation. The results of Model 1, which include a variable for parent nativity, show that parent nativity is a significant variable affecting special education participation, even when controls for low income, gender, race/ethnicity, limited English proficiency, grade level, and Section 504 are included. The results indicate that children of immigrants were significantly less likely than children of US-born parents to participate in special education. According to the odds ratios, the odds that a child of immigrant parents participates in special education were 62 percent that of children of US-born parents.

Table 2. Logistic Regression Analysis of Participation in Special Education, October 2013

Variable	Model 1		Model 2	
	Odds Ratio	95% CI	Odds Ratio	95% CI
Immigrant Parents[a] (Yes vs. No)	0.62***	[0.60, 0.65]		
Low Income (Yes vs. No)	1.82***	[1.78, 1.87]	1.80***	[1.76, 1.84]
Gender (Male vs. Female)	2.29***	[2.25, 2.34]	2.29***	[1.76, 1.84]
Race/Ethnicity[b]				
Hispanic (any race)	1.00	[0.97, 1.03]	1.06***	[1.03, 1.09]
Black (non-Hispanic)	1.06**	[1.02, 1.10]	1.02	[0.98, 1.05]
Asian (non-Hispanic)	0.58***	[0.56, 0.61]	0.49***	[0.46, 0.51]
Multiracial (non-Hispanic)	0.92**	[0.88, 0.97]	0.93**	[0.89, 0.98]
Native American (non-Hispanic)	1.09	[0.92, 1.30]	1.05	[0.89, 1.25]
Pacific Islander (non-Hispanic)	0.88	[0.66, 1.17]	0.83	[0.63, 1.10]
Limited English Proficiency (Yes vs. No)	0.99	[0.95, 1.02]	0.83***	[0.81, 0.86]
Grade Level	1.15***	[1.14, 1.15]	1.14***	[1.14, 1.15]
Section 504 (Yes vs. No)	0.49***	[0.46, 0.52]	0.49***	[0.46, 0.52]

[a] Students whose first language is other than English or Spanish are identified as children of immigrants.

[b] The reference group for race/ethnicity is white (non-Hispanic)

*p <= 0.05; **p <= 0.01; ***p <= 0.001

Comparison of Models 1 and 2 suggests the impact on the other covariates of omitting the variable for parent nativity. The direction and significance of odds ratios for several variables do not change: low income (significant positive), male gender (significant positive), Asian and multiracial vs. white (significant negative), Native American and Pacific Islander vs. white (insignificant), grade level (significant positive), and participation in Section 504 (significant negative). However, when a variable for parent nativity is added to the analysis, an insignificant relationship between being black (vs. white) and receiving special education becomes significant and positive, and a significant negative impact of English language proficiency on receiving special education loses significance. (The definition of the proxy for having an immigrant parent makes changes in the coefficient on Hispanic/Latino ethnicity difficult to interpret.)

Special Education Placement

Table 3 details the results of the analysis of placement into substantially separate settings, and Table 4 details results of the analysis of placement into general classrooms at least 80 percent of the time. The multivariate analyses show that even when controlling for other factors that have been correlated with placement, parent nativity may in fact be a significant factor in how restrictive the student's placement is. Among students receiving special education, children of immigrants were significantly more likely than children of US-born parents to be educated in substantially separate settings (Table 3) and significantly less likely to be included in general education classes at least 80 percent of the time (Table 4). The odds that children of immigrants were in a substantially separate setting was 124 percent the odds for children of US-born parents, and the odds that children of immigrants were included in general classes at least 80 percent of the time was 77 percent that of children of US-born parents.

Table 3. Logistic Regression Analysis of Special Education Students Placed in Substantially Separate Settings, October 2013

Variable	**Model 1**		**Model 2**	
	Odds Ratio	95% CI	Odds Ratio	95% CI
Immigrant Parents[a] (Yes vs. No)	1.25**	[1.10, 1.41]		
Low Income (Yes vs. No)	1.63***	[1.52, 1.75]	1.64***	[1.53, 1.76]
Gender (Male vs. Female)	1.12***	[1.05, 1.19]	1.12***	[1.06, 1.19]
Race/Ethnicity[b]				
Hispanic (any race)	1.45***	[1.33, 1.57]	1.42***	[1.30, 1.54]
Black (non-Hispanic)	1.52***	[1.38, 1.68]	1.55***	[1.41,1.71]
Asian (non-Hispanic)	1.57***	[1.33, 1.85]	1.73***	[1.49, 2.02]
Multiracial (non-Hispanic)	1.07	[0.93, 1.25]	1.08	[0.93, 1.25]
Native American (non-Hispanic)	1.74*	[1.04, 2.90]	1.76*	[1.05, 2.94]
Pacific Islander (non-Hispanic)	2.58*	[1.17, 5.69]	2.66*	[1.21, 5.82]
Limited English Proficiency (Yes vs. No)	0.74***	[0.67, 0.81]	0.78***	[0.71, 0.86]
Grade Level (K-5)	1.02	[1.0, 1.04]	1.02	[1.00, 1.04]
Section 504	0.59***	[0.46, 0.75]	0.59***	[0.46, 0.75]
Primary Disability[c]				
Communication	1.15*	[1.02, 1.29]	1.15*	[1.02, 1.29]
Developmental Delay	3.32***	[2.97, 3.73]	3.33***	[3.00, 3.73]
Autism	13.79***	[12.34, 15.42]	13.81***	[12.35, 15.44]
Health	1.76***	[1.53, 2.02]	1.75***	[1.52, 2.02]
Emotional	8.40***	[7.43, 9.50]	8.38***	[7.41, 9.48]
Neurological	5.09***	[4.36, 5.95]	5.09***	[4.36, 5.95]
Intellectual	26.97***	[23.44, 31.03]	27.05***	[23.51, 31.12]
Multiple	11.27***	[9.52, 13.35]	11.28***	[9.52, 13.53]
Physical	1.23	[0.90, 1.68]	1.23	[0.90, 1.69]
Deaf	2.06***	[1.46, 2.93]	2.11***	[1.49, 2.98]
Blind	1.69	[1.0, 2.84]	1.70*	[1.01, 2.86]
Deaf and Blind	5.24***	[2.59, 10.58]	5.42***	[2.69, 10.93]

[a] Students whose first language is other than English or Spanish are identified as children of immigrants.

[b] The reference group for race/ethnicity is white (non-Hispanic).

[c] The reference group for primary disability is specific learning disability. In this sample, students with specific learning disability have the lowest rate of substantially separate placement.

*p <= 0.05; **p <= 0.01; ***P <= 0.001

Table 4. Logistic Regression Analysis of Special Education Students in General Classroom at Least 80% of the Time, October 2013

	Model 1		Model 2	
	Odds Ratio	95% CI	Odds Ratio	95% CI
Immigrant Parents[a] (Yes v. No)	0.77***	[0.70, 0.85]		
Low Income (Yes vs. No)	0.74***	[0.70, 0.78]	0.74***	[0.70, 0.78]
Gender (Male vs. Female)	0.96	[0.92, 1.01]	0.96	[0.92, 1.01]
Race/Ethnicity[b]				
Hispanic (any race)	0.70***	[0.66, 0.75]	0.72***	[0.67, 0.77]
Black (non-Hispanic)	0.68***	[0.62, 0.73]	0.66***	[0.61, 0.72]
Asian (non-Hispanic)	0.76***	[0.67, 0.86]	0.68***	[0.61, 0.77]
Multiracial (non-Hispanic)	0.87*	[0.78, 0.98]	0.87*	[0.78, 0.98]
Native American (non-Hispanic)	0.69	[0.47, 1.01]	0.68*	[0.46, 1.00]
Pacific Islander (non-Hispanic)	0.52*	[0.27, 0.99]	0.51*	[0.26, 0.97]
Limited English Proficiency (Yes vs. No)	1.19***	[1.10, 1.29]	1.11**	[1.03, 1.19]
Grade Level (K-5)	0.89***	[0.88, 0.91]	0.89***	[0.88, 0.91]
Section 504 (Yes vs. No)	1.77***	[1.49, 2.11]	1.77***	[1.49, 2.11]
Primary Disability[c]				
Communication	0.67**	[0.53, 0.85]	0.68**	(0.54, 0.85]
Special Learning Disability	0.54***	[0.43, 0.68]	0.55***	[0.43, 0.69]
Developmental Delay	0.29***	[0.23, 0.36]	0.29***	[0.23, 0.36]
Autism	0.09***	[0.07, 0.12]	0.09***	[0.07, 0.12]
Health	0.50***	[0.39, 0.63]	0.50***	[0.39, 0.64]
Emotional	0.10***	[0.08, 0.14]	0.11***	[0.08, 0.14]
Neurological	0.22***	[0.17, 0.28]	0.22***	[0.17, 0.29]
Intellectual	0.04***	[0.03, 0.05]	0.04***	[0.03, 0.05]
Multiple	0.09***	[0.07, 0.11]	0.09***	[0.07, 0.12]
Deaf	0.13***	[0.09, 0.17]	0.12***	[0.09, 0.17]
Blind	0.31***	[0.21, 0.46]	0.31***	[0.21, 0.46]
Deaf and Blind	0.83***	[0.05, 0.14]	0.08***	[0.46, 0.14]

[a] Students whose first language is other than English or Spanish are identified as children of immigrants.

[b] The reference group for race/ethnicity is white (non-Hispanic).

[c] The reference group for primary disability is physical disability. In this sample, students with physical disability have the highest rate of participation in general classrooms at least 80% of the time.

*p <= 0.05; **p <= 0.01; ***P <= 0.001

Comparison of Models 1 and 2 in both sets of regressions indicate that introducing the proxy for having an immigrant parent does not alter the direction and significance of relationships between other independent variables and placement. In addition to children of immigrants, low-income and nonwhite students were more likely than their counterparts to be placed in substantially separate settings and less likely to be included in the general classroom at least 80 percent of the time.

Case Study Methods

The case study examines how educators in a Massachusetts public elementary school responded to difficulties experienced by 11 students who were children of immigrants. According to Yin (2009), the case study approach is preferred when the researcher has little control over events, attempts to answer explanatory how and why questions requiring in-depth investigation, and examines contemporary phenomena in a real-life context which is pertinent to the phenomena of study. The unit of analysis for this study is the school. Through examining the school's approach to the 11 students, I explore how public policy and organizational structure are influential in a class of cases in which students have parents who grew up in another country and have low income. The study generalizes to theory, rather than to a population, in a way that is generative of implications for policy and practice.

Selection of School

The selection process for the school was purposeful (Maxwell 2013). To maximize the likelihood of recruiting a sufficient number of participants meeting the study criteria within a school, I focused recruitment efforts on school districts in Massachusetts cities with the highest numbers of foreign-born, unnaturalized people with incomes below 200 percent of the federal poverty level (FPL; identified through the US Census) and with schools having higher than statewide proportions of students whose first language was not English and classified as low income/economically disadvantaged (identified through the state's DESE website). I received consent from one of these cities' superintendents to carry out the study in his district. The superintendent selected the study school from among those that met the study criteria.

The school selected for the study included pre-K through fifth grades and had in the school year of this study (2016–17) a total of 457 students and a 10:1 student to teacher ratio. It was a Title I school that housed a districtwide program for students with autism. All the teachers were licensed and considered to be highly qualified in their area of teaching. Only two of the staff, the school adjustment counselors, were fluent in a language other than English, which was Spanish. In the previous school year, the school was, based on student standardized test performance, in the 44th percentile of the state's elementary schools, with a Level 2 rating—the second highest rating on a scale of 1 to 5. The per-pupil expenditure for this school's district ($20,620) was above the statewide average of $15,911.

Almost half of the school's students were identified as Hispanic and just over a third were identified as white. Approximately half of the students had a first language other than English, including Spanish, Haitian Creole, Portuguese, and 18 other languages. Just over a quarter of the students were categorized as English language learners (ELLs). The parents of 168 (37 percent) of the students requested school communications in Spanish. An additional 13 parents requested communications in nine other non-English languages. Approximately a third of the students were categorized as economically disadvantaged, likely a low estimate because it includes only students whose families participated in a supplemental nutrition assistance program, means-tested income assistance, state foster care, or Medicaid. With the exception of state foster care, these programs are not available to or elected by all families that are poor enough to qualify for them.

Almost a quarter of the school's students received special education, a comparatively high proportion because the school housed the districtwide program for autism through which the district sought to place all of its elementary students with autism regardless of their home address. The rate of special education participation for the district as a whole, approximately 18 percent, was close to the state average of 17 percent.

Overall, and with the exception of the autism program, the school had a common, traditional structure of one general education teacher per classroom, one grade level per classroom, and student promotion to a new teacher each year. School programs for students needing additional help included a child study team, special education, Section 504 accommodation, ELL services, Title I, summer school, reading and math specialists, an adjustment counselor, and volunteer reading and big brother/big sister programs.

Selection of Focal Students, Parents, and School Staff

Criteria for selection of the 11 students in this study were the following: (1) their parents had immigrated to the United States after age 16, (2) their parents had concerns about the child's academic or social performance at school, and (3) the students were eligible for the free or reduced price lunch program, which meant that family income was within 185 percent of the FPL.

To recruit students through their parents, the school and a district family literacy program worked with me to distribute fliers to parents in English, Spanish, Haitian Creole, and Portuguese languages. I also presented the study to parents at a Special Education Parent Advisory Council meeting facilitated by a parent bilingual in Spanish and English. I held informational meetings with 13 parents, assisted by bilingual interpreters, and provided written information on the study in the language requested by each parent. In addition to communicating the purpose and details of the study, I informed parents I would not ask them about their legal status. I offered language interpretation and $20 gift cards from a local supermarket for each hour of interview time. The parents of 11 children consented to participate in the study.

Of the parents who participated in the study, four were single mothers. For three students, I interviewed both parents (mother and father); for eight students, only the mother. Parents came from six different countries. For all but one parent, Spanish was the first language, and the other parent was not fluent in English and spoke a very low-incidence language. All parents except three chose to have a language interpreter assist in the interview. The parents' own formal educational attainment varied widely. One parent had never been to school. Three had some elementary school, three had some high school, and five had graduated from high school or had a certificate of equivalence. One had some college, and one was a college graduate. The longest that any of the parents had been in the United States was 20 years. Three parents had lived in the United States for less than 10 (7 to 9) years, 8 for 10 to 15 years, and 3 between 16 and 20 years.

All 11 students were born in the United States. Five were girls; six were boys. They spanned all grades in the school, pre-K through 5. When I interviewed the parents, five of the students were receiving special education services—four of them through the school's autism program. All students had Spanish as their first language, except for one whose language was recorded by the district as English; his parent was the one who spoke the low-incidence language and struggled with expression in English. None of the students had a parent whose first language was English, and all their parents, except for one whose parent was fluent in English and another whose parent spoke the very low-incidence language, requested school communication in their language. All of the students were covered by the Massachusetts Medicaid program. Eight of the students had been in pre-K programs, and four had received early intervention services (Table 5).

Table 6 provides a summary, for each student (named with a pseudonym), of the concerns that were expressed by their parents and staff about them and the services provided at school during the study year. It also includes information on the source of referral to special education if it occurred prior to the study year.

The original study plan envisioned interviews with all school staff involved in providing educational or developmental services to the students of focus. I asked participating parents to identify school staff members who worked with their children. With the parents' consent, I sent emails to the identified staff, cc'ing the principal, requesting an interview. Often, when interviewed, staff members identified other staff working with the student of focus, of whom the parents were not aware. If I had the parent's permission, I would then contact those additional staff members (cc to the principal) to request an interview. Participating staff who did not withdraw from the study (see below) worked with nine of the focal students and included six general education teachers, three special education teachers, one ELL teacher, a math coach/specialist, and a paraprofessional. The two second-grade students, Elena and Florencia, had the same teacher, and two of the fourth-grade students, Ian and Jack, had the same general and special education teachers.

Table 5. Characteristics of Sample Students

Total # Students: 11 Born in United States: 11 **Gender** • Boys: 6 • Girls: 5 **Grade** • pre-K: 1 • K: 2 • 1: 1 • 2: 2 • 3: 2 • 4: 2 • 5: 1 **Special Education Status** • Not receiving special education: 6 • Receiving special education: 5 o Autism program: 4	**First Language** • Spanish: 10 • English + low-incidence language: 1 Parent or teacher concerned about English language proficiency: 9 Receiving ELL Services: 6 **Health Coverage** • Medicaid (MassHealth Standard): 10 • Employer Insurance and Medicaid Supplement (CommonHealth):1 Early Intervention: 4 Preschool: 8 Parents request school communication in Spanish: 9

Table 6. Summary of Parent and Staff Concerns About and School Response to Students

Pseudonym/ Grade	Parents' Concerns	Staff Concerns	School Response
Adela pre-K	Student nonverbal since age 2. Specialists tell parent student may have autism, but the diagnosis is unclear.	Student nonverbal, lacks interest in peers, has low academic skills, is a slow learner, and has autism.	Connected to special education by pediatrician, in substantially separate autism program, 15-minute monthly consultation from behaviorist, pull-out speech therapy for 30 minutes 2 times per week. Daily log for parent.
Beatriz K	Student has difficulty speaking Spanish and English, low volume of verbal communication, shyness, difficulty with math.	Student seems not to understand or speak Spanish or English, shuts down when frustrated, has unusual difficulty with math, may have a disability.	Student red-flagged for a second developmental screening, receives push-in ELL services with 8 other students and 45 minutes per day small group pull-out. Teacher brought student to child study team twice, met with one of parents, and sends graded work home to parent.
Carlos K	Retained for second year of kindergarten, does not speak English, shy, ridiculed by other students for repeating grade K.	Reading below grade level, very low English proficiency, student raises hand but doesn't speak when called on.	Student evaluated at the end of second year of kindergarten at the request of teacher. Parent did not receive evaluations in Spanish. An occupational therapy evaluation in English indicates delays in eye–hand coordination and recommends occupational therapy.

David 1	Student struggling with English language. Below grade level in reading and math.	Staff not interviewed.	Teacher asked parent to have family speak English at home. Student was offered summer school and a volunteer to read with him at lunch and recess.
Elena 2	Struggling with English language, below grade level in reading.	Struggling somewhat with reading, but more so with math.	In class with push-in ELL teacher who works with all students in class for 1.5 hours per day; with literacy specialist in small in-class group for 30 minutes 4 days per week; receives some in-class support from math coach. Parent asked to have student read each night.
Florencia 2	Struggling with English language, difficulty with reading and math, distractibility, low tolerance for frustration.	Slightly below grade level in reading, struggles with math due to low self-confidence, difficulty paying attention, almost fluent in English.	Summer school, in class with push-in ELL teacher who works with all students in class for 1.5 hours per day; with literacy specialist in small in-class group for 30 minutes 4 days per week for first half of year, receives in-class math support from math coach and Title 1 tutor, participated in lunch reading with volunteer.
Gabriel 3	Struggling with reading. Not sure if this is due to English language learning.	Below grade in reading and writing; not progressing as other students. Teacher asks whether this is due to a disability.	In classroom with ELL teacher pushing in for 45 minutes per day to work with 7 students; summer school; small group with reading specialist 3 times per week for 30 minutes. Volunteer reads with him during lunch. Teacher brings him to child study team twice.
Henry 3	Born prematurely, ADD, below grade level in reading and writing, difficulty processing information.	Autism, difficulty with fine motor skills, improving in handling frustration and peer relationships.	Connected to special education by Early Intervention, in autism program integrated classroom with 2 teachers and 2 paraprofessionals. Literacy specialist works with student in class 3 times per week for 30 minutes; behaviorist provides consultation 30 minutes per month. Social skills group once per week. Accommodations for written output. Teacher sent work home for parent to do with student.
Ian 4	Autism, dyslexia, ADHD, language-based learning disability, emotional disability, difficulty with reading, writing, social skills, impulsive behavior, heightened sensitivity to sound.	Reading at first-grade level, low English language, low social skills, impulsive behavior, frustration, anxiety.	Connected to special education by Early Intervention, in autism program substantially separate setting with a special education teacher and paraprofessionals. Attends, with paraprofessional, general education class for math, gym, music, art, drama, science, lunch, and indoor recess. Behaviorist observes student in class once a month and consults with special education teacher once a week. Adjustment counselor works with student. Speech and language therapist provides consultation to teachers and adjustment counselor. Daily log between school and home.

Jack 4	**Depression, under-stimulated at school.**	**Autism, processes slowly, lacks self-confidence, avoids challenge, difficulty with social skills.**	**Connected to special education by Early Intervention, in autism program, but fully included in general classroom. Social skills group, summer school. Has big brother volunteer through parent request.**
Karina 5	Anxiety, difficulty with math.	Specific learning disability in math, anxiety, low self-esteem, lack of motivation, poor memory, slow processing and production.	Brought to child study team, teacher referred for special education evaluation, student evaluated. IEP says student to receive 15-minute monthly consultation from special education teacher, three 30-minute per week sessions of in-class math support from special education teacher or paraprofessional, weekly 30-minute pull-out with adjustment counselor. IEP for next year includes removal from general classroom for math.

Data Collection

Data consisted of semi-structured in-depth interviews with the school principal, parents, and school staff, as well as document review. Interviews began in late March 2017, allowing parents and staff to reflect on and talk about the 2016–17 academic year. Interview topics were derived from the research questions and the theoretical frameworks guiding the study. All interviews were audio-recorded with the consent of the interviewees with the exception of one parent who requested his interviews be recorded through handwritten notes.

In May 2017, the principal, citing concern about possible legal action from the parents, withdrew the school from the study upon learning I had given Karina's parents the contact information for the special education call center of the regional parent training and information center funded by the US Department of Education. I believe the withdrawal was symptomatic of the conflicted position of the principal regarding parents' access to information about the special education program. Although the school was required to fully inform parents of special education rights and options, doing so may have been perceived as diminishing the principal's control over constrained resources. Shortly after the principal's withdrawal, five of the staff members I had interviewed also withdrew from the study. As a result, the study does not include interviews with all the staff in the research plan, and I was unable to interview any staff working with two of the focal students.

Principal Interview

I conducted two hour-long interviews with the principal in her office at the school. The interviews focused on school organizational structures; general procedures, structures, and resources for working with parents; and procedures, structures, and resources for responding to students' academic and social difficulties.

Parent Interviews

I conducted hour-long interviews with parents at a location of their choosing. The number of interviews with each parent varied from one to three as a result of differences in their availability. I met with Karina's parents several additional times at their request to help with translating school documents and locating services in the community for their daughter. Interview topics included the concerns they had about their children and their knowledge of school and community resources and programs and of the school staff who worked with their children. I also asked parents to identify and describe their interactions with school staff.

The parents of nine of the students requested a bilingual (English and Spanish) interpreter to facilitate the interview. I provided training to bilingual interpreters on the study and its ethical standards. I asked interpreters to translate, as much and as accurately as possible, the parent's exact words. Interpreters signed a confidentiality agreement in the presence of the interviewee.

Staff Interviews

I met individually with staff in the school building, either before or after school or during a planning period, for one-hour interviews on each focal student. One staff member who did not have time to meet contributed responses to the interview questions in writing. In interviews, I asked staff to describe their efforts to assess, plan, and deliver services to their focal student; identify other staff working with the student; describe their interactions with parents and other staff concerning the student; and provide their perception of influences on the capacity of school staff to meet their student's needs.

Documents

Documents used in this study included those pertaining to the school's organizational structure, procedures, and resources, generally and particularly for working with parents and responding to students having academic or social difficulty. School and district-level documents included their webpages, the district budget, the school district's elementary school handbook, student report cards, student individual education programs (IEPs), and the school's child study group referral form. State-level documents included the state procedural safeguards document for parents of children receiving special education. Documents pertaining to public policy included state and federal legislation, regulation, an executive order, and guidances pertaining to elementary education, special education, ELL education, and language interpretation and translation for parents.

Data Analysis

I transcribed handwritten interview notes and the audio-recorded English words of the interviewer, interpreters, and parents. I uploaded all interviews into Atlas.ti version 8, which I then used to code the data. I created an initial list of a priori codes from the research questions and theories used to guide the study. As I coded, I also developed inductive codes when new themes emerged. I used the constant comparative method (Corbin and Strauss 2008) to check codes and the data attached to them for conceptual consistency, making sure that properties and dimensions were consistent across incidents for each code. As I coded, I wrote memoranda pertaining to the research questions. I grouped the interviews by focal student and ran query reports for each student pertaining to each of seven categories:

1. Parent/staff concerns about the student
2. Actions taken by school staff to respond to the needs of the student
3. How the staff worked together for each student, including subcategories for each of relational coordination theory's seven dimensions of communication and relationship
4. How the staff worked with the parent, including subcategories for each of the relational coordination theory's seven dimensions of communication and relationship
5. Influential organizational structure
6. Influential policy
7. An emergent category for level of resources

I then looked across the students' data in each category to answer the research questions. I used pattern matching (Yin 2009) to compare the organizational theory with empirically derived data.

Case Study Findings

I found that public policies, district and school organizational structure, and constrained resources combined to compromise the ability of school staff to meet the needs of the study students—children of immigrants—in a timely way. In this paper, I will confine myself to discussing those findings relevant to understanding how school organizational structure and public policies affecting the educator–parent relationship contributed to the outcomes of the quantitative analysis. (Other contributing factors that emerged in the study included insufficient structure to support the coordination of staff regarding individual students; siloing of general education, special education, and ELL staff; state policy requiring English-only education; and district and school practice of delaying, for several years, special education evaluation of students suspected of having a disability when their first language is not English.)

School Structures for Educator–Parent Coordination

Interviews revealed that study parents were not able to participate effectively in meeting the academic needs of their children because the structures that shaped their relationships with staff did not provide them with necessary opportunities, knowledge, or consistent language access. Data revealed that the structures shaping staff–parent relationships were primarily bureaucratic in nature, generally placing parents in a subordinate role regarding the education of their children, as opposed to the partnership proposed by relational bureaucracy theory. There were few opportunities for parents and staff to connect about individual students and an apparent indifference to ensuring that parents were knowledgeable of their children's school experience and of options for supporting their children. Reflecting on her experience with the school in trying to meet her daughter's needs, Karina's parent noted the detachment that she felt characterized the school's relationship with her as her daughter's parent. "I imagine this must have happened to all parents with children who have special needs that—you know, they don't—the schools just—kind of just get it done, go through the motions of things, and don't really pay attention to what needs to be done."

Structures that provided information to parents on school operations and programs included back-to-school night, the school website, and the district's elementary handbook. The district elementary school handbook, printed in English as well as two of the district's most commonly spoken non-English languages, provided general information and had sections on topics of importance to parents in this study: English language learners, the Parent Information Center, special education, and the roles of the adjustment counselor and literacy specialist. However, it did not explain that parents could request a special education evaluation for their child, and it did not include any information on Section 504 accommodations, Title I tutors, or summer school programs. Additionally, the handbook had a very high readability level, varying from a 12th-grade to a graduate school reading level on the SMOG readability index.

Structures connecting school staff and parents regarding the progress of individual students included report cards and parent–teacher conferences. Report cards, issued three times a year, also had a high readability level—a college reading level on the SMOG readability index—and assumed knowledge of the concept of educational standards. Ratings were "not yet progressing toward the standard," "progressing toward the standard," "meeting the standard," and "exceeding the standard." General education teachers were required to be available for one parent–teacher conference per student per year and were given three early release days for this purpose. Although the school had a child study team to discuss students suspected of having a disability, parents were not alerted when their child was brought to the attention of the team, nor were they invited to attend a meeting. There were additional structures, mandated by special education law, that connected staff with parents of students participating in special education. These included one annual IEP/eligibility meeting with members of the school IEP team and three additional progress reports on students' achievement of IEP objectives.

There were two structures providing support to study parents to assist their integration and communication. One was the Parent Information Center (PIC), which was the district's source for translation and interpretation services. The other was a family literacy program that supported a small number of immigrant parents of preschool children, in which Adela's parent participated. When registering their children for school at the PIC, parents could indicate whether they would need to have written school materials, including report cards, translated into a non-English language, and staff could contact the PIC via the Internet for language interpreters or to have written materials translated into a non-English language. The district family literacy program provided English classes and, secondarily, information on children's schooling, to a small number of parents of children attending pre-K classes at several district schools, including the study school, and a local Head Start program.

Language Barriers

Language difference emerged as a significant barrier to staff–parent communication. Although 37 percent of the school's students had parents who had told the PIC they required school communication in a non-English language, there was no multilingual capacity—only English—in the front office. Parents and teachers of five of the study students identified language difference as a factor that contributed to delayed and too-infrequent communication with one another. Teachers said language difference reduced their use of email and telephone to communicate with the parents and explained that it could take up to two weeks to obtain an interpreter through the PIC. Parents described the time (several days) and substantial effort required to obtain the help they needed to compose notes to the teacher in English.

Language difference also affected the accuracy of communication. Adela's special education teacher explained how hard it was for her to communicate with the student's mother: "She gets picked up every day, and [the parent] is here. You can have those short exchanges. I don't understand if [the parent] knows what I'm saying, and I feel bad that I'm talking in English to [the parent] when I know it's not [the parent's] first language." The teacher used Google Translate to translate into Spanish the daily logs she sent home to the mother. She acknowledged, "It doesn't come out that great."

The PIC was inconsistent in providing the interpretation and translation that parents and staff needed to communicate with one another. Although interpreters frequently attended the parent–teacher conferences for the parents who were not fluent in English, this was not always the case. When Carlos's teacher invited his parent in to recommend that Carlos have a special education evaluation, an interpreter was not present although needed. The parent shared, "That day it was particularly difficult because I wanted to tell her more things … and to understand what she was saying." Henry's parent, who had quite limited English and spoke a very low-incidence language, never had an interpreter to assist her communication with staff. Henry's special education teacher explained, "Parent involvement has been … difficult … because the language barrier is challenging. Our communication between home and school is limited because of that."

Parents related incidents in which the quality of interpretation, when provided, was problematic. David's parent attended a presentation for her son at the back-to-school night.

> All of the teachers for the grade were there. … And the teachers went up, and they all spoke, but the [interpreter] never said anything. At the end, she asked in Spanish, "Does anyone have any questions?" … We didn't know what happened. For the parents who didn't speak English, … we left as we [had] arrived. We didn't know anything.

Karina's parent walked into her student's special education eligibility meeting without knowing what it was. About seeing nine professionals at the table, she said, "I was so nervous. … I was there by myself with so many people. I turned to the interpreter and said, 'I don't think this is a meeting where I should be by myself.'" She reported that the interpreter did not explain to the other attendees what she had said and advised her to let the meeting proceed. She reported feeling overwhelmed. She left the meeting without understanding that her daughter had a learning disability in math, that she had been found eligible for special education, or what the proposed IEP contained in terms of goals, services, accommodations, or placement.

Carlos's parent also reported her experience of not understanding the proceedings of her son's special education eligibility meeting despite the presence of an interpreter.

> It was overwhelming. … Everyone was talking to me at the same time, so the teachers were talking to me, and then the other ones were talking to me while the interpreter was talking to me. So, it was just a lot of information at once. … It was too fast. … I would begin answering one, and she [the interpreter] would be moving towards … the fourth question. … I was so overwhelmed. … They … told me in one hit, he's going to need help, he has a problem. … I left that meeting on the verge of tears. … To hear that your child needs help is something devastating. … I was crying, they told me my son has issues … and at the end, I really don't even know what the issue is. … I was very confused.

This parent, like Karina's, left the meeting without an understanding of what had taken place.

The translation of written materials was also inconsistent, even in Spanish, the most common non-English language at the school. Carlos's and Florencia's parents never received a report card in Spanish although they had registered their need for translated material with the PIC. Carlos's and Karina's parents did not receive translation of their children's special education evaluation reports until months after their initial eligibility meetings. Carlos's parent explained that translated school documents often contained important mistakes. The proposed IEP for Karina was not accurately translated. "Paraprofessional" was translated into Spanish as "professional," its opposite in meaning, each time it appeared in Karina's IEP.

Extent to Which Knowledge of Student Shared

Parents had very limited knowledge of who was working with their children, their roles and qualifications, and what services their children were receiving in school. All parents knew of their child's general education teacher, but none of the parents knew of all of the staff who worked with their child at school. The parents of Beatriz, Carlos, David, Elena, Florencia, and Gabriel all had concerns about their child's English proficiency, but not one of those parents knew there was an ELL teacher working with their child or how language was being addressed with their child at school. The literacy specialist worked with Elena, Florencia, Gabriel, and Karina, but three of the parents had no awareness of the work she did with their child, and one thought she was an aide. Although the math coach worked with Elena, Florencia, and Karina, their parents, who all reported concerns about math, did not know about it.

Although the parents of children participating in special education had, because of legal requirements, been exposed to a broader range of staff in the special education evaluation and planning process, they had similar holes in their knowledge of staff and services for their child. Of the parents of the four study students who received special education evaluations (Beatriz, Carlos, Henry, Karina), none knew the professions of the people performing the evaluations. The parents of the three students who had subsequent special education eligibility meetings (Carlos, Henry, Karina) did not know the roles of several of the people attending the meeting. All parents whose children were in the autism program knew of the involvement and role of their child's special education teacher. However, the parent whose child received special education outside of the autism program (Karina) did not know there was a special education teacher who was supposed to be providing services to Karina. The parents of three of the students receiving special education with whom paraprofessionals were working (Adela, Henry, Karina) did not know what a paraprofessional was and how a paraprofessional differed from a credentialed teacher. Adela's parent thought that the education team leader supervised the teaching staff that worked with her daughter, although her role was confined to coordinating special education evaluations and meetings. Jack's parent, concerned about his son's emotional well-being, thought the person administering his social skills group was an aide, although she was a certified behaviorist.

Interviews revealed that parents lacked key knowledge of how their child was doing at school and often had a view of the student that differed from staff's.

- The parents of Beatriz, Carlos, Gabriel, and Karina had all not been aware of the depth of concern the staff had about their child or that their teachers had long suspected that their child might have a disability.
- Carlos's parent did not know that an occupational therapy evaluation had found her son to have delays in eye–hand coordination.
- Karina's parents did not know their daughter had been found to have a specific learning disability in math.
- Adela's special education teacher did not know outside clinicians questioned whether she had autism, and Adela's parent was not aware the teacher perceived Adela as having low academic skills and a slow learning profile.
- The teacher of Elena and Florencia expressed less concern about those students than did their parents. Florencia's mother believed the teacher's expectations for her daughter were too low.
- Henry's parent believed he had ADD and delays stemming from premature birth, while his teachers believed he had autism and had no knowledge of an ADD diagnosis.
- Ian's parent was not aware that his son's special education teacher believed some of his academic problems stemmed from limited English. She did not raise concerns about his English. Although his teachers shared the mother's concern about Ian's significant difficulty with reading, they did not talk about his dyslexia, as had the mother.
- Jack's parent was told by a clinician outside of the school that Jack did not have autism and, instead, suffered from depression. However, his teachers believed he had autism. While Jack's mother believed he needed more stimulation at school, particularly in math, his teachers said he shied away from academic challenges. While the mother believed her son was depressed, the teacher said he was not sad. Jack's teachers had concerns about his peer relationships, particularly his ability to stand up for himself, of which the mother seemed unaware.

Difficulty in sharing knowledge about the students was bidirectional. Staff tended to have limited knowledge of the parent's work on behalf of the student. Adela's, Ian's, and Jack's parents were getting rich diagnostic information on their children and therapeutic services for them from developmental clinics independent of the school district. Adela's special education teacher did not know the types of clinicians Adela was seeing or what they were finding. Ian's parent thought the school staff had sufficient knowledge of her work with him outside of the school, but the general and special education teachers who worked with Ian said they did not. Jack's parent said that school staff had not asked her what she was doing to support her son outside of school.

Systems Knowledge

Parents had very little knowledge of the programs that could be accessed to assist their children at school. When first interviewed for this study, the parents of Beatriz and Karina did not know of the existence of summer school programs for which they would have been eligible. No parent knew there was a process for requesting an alternative to sheltered English immersion instruction for ELL students with special individual needs (see next section). None of the parents knew of the child study team or the Section 504 program, and, with the exception of the parents of the four students in the autism program, none of the parents of the other seven students knew of the special education program. Additionally, none of the parents knew that they could request an evaluation of their child for eligibility for special education. No knowledge of the special education program was shared by parents whose children had been or were in the process of being evaluated (Beatriz and Carlos), including Karina's parents, even months after Karina had actually formally begun to receive special education services.

None of the study parents whose children had received special education evaluations knew that they could play a role in deciding which evaluations to provide or that they could request an independent evaluation. The parents of Carlos and Karina, who were evaluated for special education during the study year, did not understand that they had been asked to attend or that they had attended a special education eligibility/IEP meeting. They did not understand the role they could play in deciding on eligibility or shaping an IEP, did not know they were entitled to receive all evaluation reports and IEPs in their first language within a prescribed time frame, or that they could accept, reject, or partially reject an IEP. They did not know of their child's right to a least restrictive placement. They did not understand the significance of their signature on the IEP form or their procedural rights. Additionally, Jack's and Karina's parents, who both had concerns about their children's emotional well-being, expressed that they did not know how to access school- or community-based mental health services.

Also emerging during interviews was the fact that parents' lack of systems knowledge could be as basic as not knowing what a superintendent was or what that role entailed. Additionally, emerging as an important aspect of missing parent information was cultural capital, or an understanding of the behaviors that school staff would find acceptable on the part of parents. For example, Karina's parents received a written notice from the school requesting their attendance at a meeting about their daughter. Although they did not understand that this was a special education eligibility/IEP meeting or what that was, they perceived the importance of the meeting from the nature of the form and were distressed that the father was unable to take the time from work to attend with the mother. However, it did not occur to them to request an alternative time. When asked why, the father explained that they did not realize this was acceptable and, in fact, had been taught by other experience within the United States not to request changes in appointments. He said, "Like for example, an appointment with immigration, when they tell you, you need to be there, you have to find a way to get there. There is not much flexibility."

The parents in this study reported that no one checked in with them to make sure they understood report cards, contents of the handbook, or special education processes and documents, including evaluation reports, proposed IEPs, meetings, or the procedural safeguards document. This was so even within the context of highly stressful special education eligibility meetings in which parents were told for the first time that their child had a disability.

Public Policies

Several public policies emerged as having an important impact in shaping the school's response to the study students: immigrant integration policy, public policy concerning language interpretation and translation for non-English speaking parents, English-only education for students, early intervention, and Medicaid. The policies most directly affecting educator–parent coordination included immigrant integration policy and public policy concerning language interpretation and translation of non-English speaking parents.

Immigrant Integration Policy

The footprint of weak immigrant integration policy is evident in this study's data. Parents lacked critical information on children's educational rights, how their school district and school was organized, and on the services and programs in their schools and communities that could be leveraged to help their children succeed in school. Insufficient public policy to provide systems knowledge to immigrant parents in the areas of children's educational and developmental services created significant risk for students who were struggling because it severely challenged parents' ability to navigate and advocate on behalf of their child at school.

Language Interpretation and Translation for Parents

Public policy concerning language interpretation and interpretation for parents with limited English emerged as critical to the frequency, timeliness, and accuracy of communication and knowledge sharing between staff and study parents. Current federal policy requires public schools to provide information to parents with limited English in their native language. This policy, required for all students, is amplified by federal and Massachusetts special education law for parents of students receiving special education. As noted above, this study revealed very inconsistent and weak implementation of these public policies.

Impact of Organizational Structure on the Implementation of Special Education Law

The study indicates that school organizational structures impacted the implementation of special education and ELL policies such that study students and their parents were deprived of rights to which they were entitled. This paper narrows its focus to the impact of the school and district's organizational structure on the administration of federal and Massachusetts special education law to study students.

The absence of organizational structures to adequately support the coordination of staff and parents, and the child study team (one of the sole structures through which the staff came together to focus on the needs of individual students) resulted in poor implementation of several provisions of special education law to study students, including Child Find, parent participation in service planning, and due process.

The Child Find provision of IDEA requires state and local education agencies to ensure that "all children with disabilities residing in the State … who are in need of special education and related services, are identified, located, and evaluated." The legislation also provides that "a parent of a child … may initiate a request for an initial evaluation to determine if the child has a disability." This aspect of Child Find was not adequately administered to study students. The parents of Beatriz, Carlos, David, Elena, Florencia, Gabriel, and Karina did not know of the existence of the special education program or that they could request an eligibility evaluation of their child. The parents of Beatriz, Carlos, Gabriel, and Karina did not fully understand staff's perceptions of the difficulties their children were having at school, and they did not know their children's teachers were concerned that their child might have a disability.

The parents of Carlos and Karina expressed great distress when they learned they could have referred their children for special education evaluation at any time. At the end of Carlos's first year of kindergarten, his parent received a request from the teacher to consider retaining him for a second year. The parent had not known, during the school year, that her son was experiencing such difficulty, and she struggled to understand and discuss with the teacher whether retention would be the best option. She reported that the teacher did not tell her about special education or the option to have Carlos evaluated. She said that had she known, she would have asked to have him evaluated to understand whether promotion to first grade with supports would have been better for him than a second unsupported year in kindergarten. She was particularly distressed because she was then asked by the subsequent teacher, at the end of his second year of kindergarten, during which he again did very poorly, to have him evaluated. She had seen her son experience pain during his second year because he was ridiculed by other students for repeating. Similarly, after three years of seeing her daughter struggle with school, Karina's mother wept when she learned (during this study) that she could have requested a special education evaluation for Karina at any time.

In addition to providing parents with the right to refer their child for an evaluation of their eligibility for special education, federal and state special education law require substantial parent involvement in the process of eligibility determination. Special education policy provides that special education evaluations cannot be conducted without the "informed consent" of the parent, which is specifically defined in the federal statute as entailing "understanding." School districts are required to provide parents with opportunity to consult on the contents of proposed evaluations and which evaluators are used. Further, upon parent request, districts are required to provide parents with assessment summaries at least two days before a team meeting. Although the parents of Beatriz, Carlos, and Karina signed forms agreeing to have their children evaluated, they reported a lack of understanding of this process. Karina's parents reported they were not asked if Karina could be evaluated; they believed they were "told" Karina would be evaluated. The parents of Carlos and Karina were not offered assessment summaries prior to their eligibility meetings. None of the study parents, at the time they were interviewed, knew they could request an independent evaluation. Karina's parents learned of this right when, during the interview, they asked for help in understanding the procedural rights document the school had given them. When they later submitted a formal request for an independent evaluation, they understood the principal as telling them the school was checking the affordability of the request. After three months had elapsed, the parents reported they had heard nothing further from the school.

Special education law also requires that parents provide "informed consent" before special education and related services are provided to their child, and it requires that parents are part of the team that develops the IEP. Parents may reject or accept any part(s) of proposed IEPs for their children. However, Karina's parents had not understood the contents of the team meeting, the assessments, and the contents of their child's proposed IEP, and they did not know they could fully or partially reject or accept it. Additionally, although special education law requires that students receive services in least restrictive settings and IEP teams must justify any removal of students from the general classroom for services, Karina's parents did not understand that the IEP proposed to remove Karina from the regular classroom for math. Karina's parents signed the IEP without realizing what they were signing when the signature page was sent home in their daughter's backpack as a single sheet with a request to affix a signature and send it back.

Finally, federal special education law requires that parents receive a procedural safeguards notice that includes information on the entitlement to independent evaluations, prior written notice, parental consent, parent access to education records, and parents' opportunity to present and resolve complaints and seek administrative and legal intervention. Although Karina's parents had received from the school a copy of the state's procedural safeguards notice in their native language, it was completely inaccessible to them because of the technical, legalistic nature of the language; dense format; and high readability level—a grade 15 readability level according to the SMOG readability index. They had no understanding of its contents, and they were not aware that anyone had explained its contents to them or checked for their understanding.

Constrained Resources

Though not an a priori construct in this study, constrained resources emerged as an additional factor that interacted with organizational structure in shaping the response of school staff to the students and affecting its implementation of special education and ELL policy and federal policy on language access for non-English speaking parents. The inconsistent availability and quality of language interpretation and translation services for parents affected study students by depressing the frequency, timeliness, and accuracy of staff–parent communication and, consequently, the degree to which information on students' progress and school programs was shared.

Other areas of constrained resources that emerged in the data included the lack of time staff had to communicate with parents and staffing shortages in special education and the autism program that resulted in failure to implement student IEPs. Teachers perceived they were unable to provide the level of service needed by study students. Students with special needs would have been affected by these constraints regardless of the nativity of their parents. However, study parents were less likely, given their lack of systems knowledge and language access, to know of these deficiencies and how to address them, and, in fact, they were not aware of these problems.

The special and general education teachers who co-taught Henry's third-grade integrated class for students with and without autism explained that the school did not have the staff resources to extend this model for an additional year or to the fourth grade. This meant that three of the students in the third-grade autism class who had been integrated were to return to a substantially separate setting, and Henry, who the teachers felt had been thriving in the integrated class, would also have to return to a more restricted placement in which there would be less continuity and increased fragmentation in his school day. Thus, limited resources may not have allowed delivery of services to Henry in the least restrictive setting consistent with his need, as mandated by special education law. However, his mother was not aware of this entitlement or of the impending change.

Parent Adaptation Experience

All study parents had grown up in another country and lacked familiarity with US school and other support systems. In addition to the unavailability of systems information and inadequate translation and interpretation services, parents' limited social and cultural capital contributed to the challenges they faced in mobilizing for their children. Some parents expressed a lack of clarity concerning acceptable behavior on the part of parents at school, particularly when it came to proactive, assertive, or questioning interaction with school staff. Others had limited social networks. For some, key family members to whom they turned to for support with parenting remained in the country of origin. No parent had any person in their social network who had navigated for a child with special needs and could provide the specific knowledge they needed to navigate for their child at school. In addition, parents expressed feeling disconnected from, or devalued by, others in the school community as a result of differences in language, race, ethnicity, national origin, social class, and educational experience.

All parents participating in this study were taking some form of purposive action to address the difficulty their child experienced at school. Their participation in this study was one such action that they undertook to learn more about what could be done to help their child improve in school. All parents were motivated by deep emotional commitment to their child. They all placed a high value on education and wanted their child to succeed in school. Regardless of the degree of difficulty their child experienced, all had positive short- and long-term goals and dreams for their children. While the level of intensity of concern and engagement was indistinguishable across the parents, the parents were at varied levels with respect to the amount of systems knowledge they had acquired and the degree to which they took a lead in advocating for their child. Their children's experience of difficulty in school gave these parents an urgency to learn and adapt to complex facets of educational and developmental policies and programs affecting their children's school experience. Despite their desire to help their child, these parents did not receive adequate support from the school in understanding how their child was progressing, their child's context at school, and the options that could be drawn upon to help their children. The parents persisted in their efforts to help their children despite these obstacles. Not receiving assistance from school staff with this process was problematic. It extended the length of time of the parents' learning processes when their children needed supports put in place without delay to avoid experiences of repeated school failure, or being in a less than optimal situation over an extended period of time.

Discussion

Together, results of the quantitative analysis and case study highlight the importance of educator coproduction with parents. While the quantitative analysis strongly suggests that young students have reduced access to special education entitlements if they have an immigrant parent, the case study demonstrates that this is in part due to public policy and organizational structure that do not adequately support immigrant parents in coproducing with educators on behalf of their children.

As a result of inconsistent availability and quality of interpretation and translation services and the absence of programs to build and ensure newcomer parents' understanding of school practice and systems, parents in the study did not always fully understand how their children were doing in school, that there was a special education program, that staff questioned whether their child might have a disability, or that they, as parents, could request a special education evaluation for their child. They did not understand that school staff, who misunderstood and violated special education law, delayed special education evaluation of their children to see whether increased English language proficiency, gained over time, would remove the difficulty the child experienced.

The present study corroborated previous studies in finding that immigrant parents whose children were receiving special education were unable to participate in the IEP process in which placement was decided because they lacked knowledge or language access (Cho and Gannotti 2005; Jegatheesan 2009; Lo 2008; Park et al. 2001; Salas 2004; Wathum-Ocama and Rose 2002). Like parents in these other studies, parents who participated in the present study were not supported to understand the IEP process and did not understand the content of IEP meetings or assessments. They did not know that there were different environments in which a service could be provided or that their child had a right to a least restrictive placement. They did not know they could play a role in choosing the place where their child's services would be provided or that they could reject any part of the IEP they did not feel was right for their child. Parents who grew up in the United States and speak English are less likely to be so unempowered in the placement process.

Implications for Educators

The results of this study suggest that schools should acknowledge the impact that parent coproduction has on service quality and student outcomes, value and make possible relational coproduction between staff and parents, and take proactive steps to ensure that all parents can equitably participate in coproduction. In the case of immigrant parents, this requires, at a minimum, providing the language access and systems knowledge the parents need to participate on behalf of their children. Programs to assist with building the social and cultural capital of immigrant parents would also be helpful.

This study confirms Jakobsen and Anderson's (2013) theory of the distributional effects of coproduction that states that in contexts in which clients play an impactful role in shaping the quality and outcomes of services, such as education, there is a danger that unequal resources for participation among clients will, if not remedied, have the effect of widening disparities among those served. To avoid this widening of disparities in education, which is sensitive to the quality of coproduction, it is important to analyze the impact of coproduction on service quality and outcomes, understand the competencies and resources required for effective coproduction, look for differences in the capacity to coproduce among those served that could result in inequitable service quality and outcomes, and address these inequities by building the capacity of coproducers who are under-resourced to participate.

References

Alba, R.D., and Nee, V. (2003). Remaking the American Mainstream: Assimilation and Contemporary Integration. Cambridge, MA: Harvard University Press.

Bronfenbrenner, U. (1979). The Ecology of Human Development. Cambridge, MA: Harvard University Press.

Child Trends. (2015). Individualized Education Plans. Retrieved from http://www.childtrends.org/databank/indicators-by-topic-area/education/

Child Trends. (2018). Immigrant Children. Retrieved from Washington, DC. https://bit.ly/33bEBvG

Cho, S.-J., and Gannotti, M. E. (2005). Korean-American Mothers' Perception of Professional Support in Early Intervention and Special Education Programs. Journal of Policy and Practice in Intellectual Disabilities, 2(1), 1-9.

Coll, C. G., and Marks, A. K. (2009). Immigrant Stories, Ethnicity and Acadmics in Middle Childhood. New York: Oxford University Press.

Conger, D., and Grigorenko, E. L. (2010). Special Educational Needs of Children in Immmigrant Families. In E. Grigorenko and R. Takanishi (Eds.), Immigration, Diversity, and Education. NY: Routledge.

Corbin, J., and Strauss, A. (2008). Basics of Qualitative Research (3 ed.). Los Angeles: Sage Publications.

Eccles, J. S. (1999). The Development of Children Ages 6 to 14. The Future of Children, 9(2).

Fortuny, K., Hernandez, D., and Chaudry, A. (2010). Young Children of Immigrants, The Leading Edge of America's Future. Retrieved from Washington, D.C.:

Gittell, J. H. (2003). TheSsouthwest Airlines Way: Using the Power of Relationships to Achieve High Performance. New York: McGraw-Hill.

Gittell, J. H. (2006). Relational Coordination: Coordinating Work Through Relationships of Shared Goals, Shared Knowledge and Mutual Respect. In O. Kyriakidou and M. Ozbilgin (Eds.), Relational Perspectives in Organizational Studies (pp. 74-94). Cheltenham, UK: Edward Elgar Publishers.

Gittell, J. H., and Douglass, A. (2012). Relational Bureaucracy: Structuring Reciprocal Relationships into Roles. Academy of Management Review, 37(4), 709-733.

Hehir, T., Grindal, T., and Eidelman, H. (2012). Review of Special Education in the Commonwealth of Massachusetts. Boston, MA: Thomas Hehir and Associates.

Hehir, T., Schifter, L., Grindal, T., Ng, M., and Eidelman, H. (2014). Review of Special Education in the Commonwealth of Massachusetts: A Synthesis Report. Boston, MA: Thomas Hehir and Associates.

Hibel, J., and Jasper, A. D. (2012). Delayed Special Education Placement for Learning Disabilities Among Children of Immigrants. Social Forces, 91(2), 503-529.

Huston, A. C., and Ripke, M. N. (2006). Middle Childhood: Contexts of Development. In A. C. Huston and M. N. Ripke (Eds.), Developmental Contexts in Middle Childhood. New York: Cambridge University Press.

Jakobsen, M., and Andersen, S. C. (2013). Coproduction and Equity in Public Service Delivery. Public Administration Review, 73(5), 704-713.

Jegatheesan, B. (2009). Cross-cultural issues in Parent-Professional Interactions: A Qualitative Study of Perceptions of Asian American Mothers of Children with Developmental Disabilities. Research and Practice for Persons with Severe Disabilities, 34(3-4), 123-136.

Lo, L. (2008). Chinese Families' Level of Participation and Experiences in IEP Meetings. Preventing School Failure, 53(1), 21-27.

Park, J., Turnbull, A. P., and Park, H.-S. (2001). Quality of Partnerships in Service Provision for Korean American Parents of Children with Disabilities: A Qualitative Inquiry. Journal of the Association for Persons with Severe Handicaps, 26(3), 158-170. doi:10.2511/rpsd.26.3.158

Salas, L. (2004). Individualized Educational Plan (IEP) meetings and Mexican American Parents: Let's Talk About It. Journal of Latinos and Education, 3(3), 181-192.

Wathum-Ocama, J. C., and Rose, S. (2002). Hmong Immigrants' Views on the Education of their Deaf and Hard of Hearing Children. American Annals of the Deaf, 147(3), 44-53.

Welsh, M., Parke, R. D., Widaman, K., and O'Neil, R. (2001). Linkages Between Children's Social and Academic Completence: A Longitudinal Analysis. Journal of School Psychology, 39, 463-482.

Yin, R. K. (2009). Case Study Research: Design and Methods (Fourth ed. Vol. 5). Thousand Oaks: Sage Publications.

Impacts of Policy Changes on Worker Safety in New York City Construction Sites

MARIA FIGUEROA
Cornell University

Introduction

The ongoing construction boom in New York City has been accompanied by a dramatic surge in injuries and fatalities, with 38 percent of all occupational fatalities in 2016 occurring in the construction industry (New York Building Congress 2017; Obernauer 2018). Day laborers in the residential sector are uniquely vulnerable to workplace safety violations, with Latinos experiencing a fatality rate almost double that of all construction workers (Obernauer 2018; Rathod 2016). A growing concern related to the non-union construction sector, and a key barrier to progress in the field of workplace safety, is the lack of access to training and labor education that immigrant construction workers experience. Ninety-three percent of construction fatalities in New York City in 2017 were non-union workers (Obernauer 2019).

Factors contributing to the high vulnerability of day laborers include the minimal union presence in residential construction and the highly fragmented structure of this sector, which comprises large numbers of small contractors and often short-lived companies. The lack of union representation and industry fragmentation led to a high degree of informality in employment relations, seriously challenging the implementation and enforcement of labor and employment laws, including workplace safety laws (Bernhardt et al. 2009; Eastern Research Group 2014). Ongoing research has identified 43 informal day-labor recruitment sites across New York City, with an estimated 3,500 to 4,000 day laborers in total seeking employment at these sites (Cornell Worker Institute 2016). Previous studies estimated the total number of day laborers in the New York metropolitan area at around 10,000, including workers at informal hiring sites and those who find jobs by other means (Valenzuela et al. 2006). Recent estimates place the number of immigrant construction workers at above 60,000 for New York City alone (Gonzalez 2016).

With the goal of abating injury and fatality rates among vulnerable workers in the construction industry and in response to growing demands on the part of labor unions for increased safety measures, the City of New York enacted Local Law 196 in October 2017. This law imposes increased safety training requirements for the issuance and renewal of building permits and seeks to promote access to safety training for the target population, who include mostly immigrant day laborers and other at-risk workers. However, the timetable for the implementation of the new law experienced significant roadblocks related to both access and training capacity.

Building on existing research on benefits of safety training for US-born and immigrant construction workers, both union and non-union (Sokas et al. 2009; Evanoff and Kaskutas 2013; Dong and Platner 2004; Williams 2010), the current study assessed the impact of Local Law 196 on safety outcomes in the construction industry in New York City, with a focus on vulnerable workers. It analyzed the implementation of Local Law 196 to identify strategies or policy measures that would more effectively achieve desired outcomes and to determine whether any content or delivery changes would be needed to increase the effectiveness of the newly required safety program.

Objectives

The goal of this study was to evaluate the implementation process of Local Law 196, focusing on outcomes in two main areas: construction workplace safety and access to safety training for the target population (vulnerable and immigrant workers). The study had three specific aims. First, to examine the implementation of Local Law 196 to identify success factors for and barriers to increasing access to site-safety training for immigrant workers. Achieving this objective required conducting interviews of key stakeholders, focus group sessions with a sample of the target population, and review of relevant documents and statutes. Second, to assess the effectiveness of the newly required training program in addressing the specific needs of the target population and determine whether any changes in content, delivery approach, and outreach would be needed to improve safety outcomes. Achieving this objective required conducting focus group sessions with a sample of the target population of immigrant or non-union workers receiving services from three local worker centers. Third, to estimate whether the implementation of Local Law 196 has contributed to a significant decline in the incidence of injury rates in the local construction industry. To achieve this objective, quantitative data on program outputs and outcomes were obtained from government data sets and reports—including the New York City Department of Buildings (NYC DOB) reports—and from worker centers delivering the required training.

Key Findings

Injury rates were consistently lower than what they would have been without the implementation of Local Law 196. However, this downward trend can be only partially attributable to the new policy, as day-laborer organizations had increased their training capacity since 2016 and local prosecutors had been strengthening enforcement of workplace safety violations. For this observed decline in injury rates to be sustained, shifts in program design and enforcement approaches might be needed, including increased input from and collaboration with organizations that represent the program's target population.

Increased access to safety training for vulnerable and immigrant workers contributed to increased safety awareness, ability to prevent accidents, and a sense of empowerment for individual workers in the target population. However, there are serious limitations to safety training in the absence of other injury prevention mechanisms such as effective monitoring and enforcement of worksite safety regulations, and promotion of a safety culture among construction managers.

Significant barriers kept the target population from fully accessing the program, including limited outreach to workers and employers about the new requirements and training delivery methods that do not address specific characteristics of vulnerable and immigrant workers, including language, literacy, and fear of employer retaliation.

The implementation of Local Law 196's safety program rests primarily on the assumption that increased training and enforcement capacities would lead to better worksite safety outcomes. However, the structure and the processes put in place for implementation might be creating unanticipated consequences, such as employer practices that prioritize holding training certification cards over compliance with site-safety regulations and worker protections.

Additional measures are necessary to improve outcomes. One such measure would be to increase collaboration between the NYC DOB (the agency administering the program) and agencies with experience with labor standards (e.g., New York State Department of Labor). Another would be to provide workplace safety training to the building inspectors who are in charge of monitoring compliance with Local Law 196, such that they are able to identify safety issues, not just require the holding of safety cards. Lastly, increased input from day-labor centers in the design and implementation of policy and programs would significantly enhance outcomes and impact, as these organizations have vast experience in advocacy and provision of training for vulnerable and immigrant workers.

Methods

This study utilized quantitative and qualitative strategies to assess policy outputs (e.g., training delivery) and outcomes (e.g., changes in injury rates), as well as to capture workers' and advocates' perceptions of the effectiveness of training programs (Creswell 2014; Winter 2006). The study also involved conceptual frameworks that underscore the importance of stakeholder coalitions (e.g., public–private partnerships) for successful implementation and enforcement (Koonse, Dietz, and Bernhardt 2015; Jacobson and Wasserman 1997).

Data Collection

This research collected both quantitative and qualitative data. The quantitative data included construction injury statistics, which were used to estimate likely impacts of the policy change on the target population. The qualitative data included perceptions about the effectiveness of the safety training programs and the implementation process. The quantitative data included published statistics from the NYC DOB's monthly and annual summary reports on construction-related injuries and fatalities, as well as current employment statistics on local construction employment published by the New York State Department of Labor.

The method for collecting qualitative data on the safety training programs involved the use of convenience and purposive sampling methods to conduct four focus group sessions of non-union and largely immigrant workers who receive legal services and safety training at three local workers centers in New York City. Each focus group included between 16 and 21 participants to achieve a sample size of 82 participants (n = 82). This sample size is adequate for focus group research, given the estimated size of the target population of 3,500 immigrant construction workers potentially receiving services from local worker centers (Cornell Worker Institute 2016) and given the likelihood of reaching a 90 percent data saturation point with an average of 4.3 focus groups of eight to ten participants (Guest et al. 2016). The focus group questionnaire was designed to capture information about the adequacy of the training requirements for meeting the needs of the target population, as well as about the effectiveness of the outreach and delivery systems. The questions probed into worker knowledge gain, changes in worker attitudes and self-reported behavior related to workplace safety, injury risk reduction, and barriers to access, as well as improvements needed in safety training content and enforcement. All of the focus groups were conducted in Spanish and English by researchers who are bilingual and native Spanish speakers. At the beginning of each focus group session, the researcher(s) explained the goals of the study to participants, reviewed their rights and potential risks from participating in the study (as outlined in an oral consent form), and obtained oral consent from workers to participate and be recorded during the session.

Semi-structured interviews were conducted of worker center leaders, government agency representatives in charge of implementing the new policy, and selected members of a joint public–private advisory task force (n = 6). These interviews obtained factual information and perspectives on the policy formation and implementation processes, as well as about the effectiveness of training programs.

Methods of Analysis

Analysis of qualitative data involved synthesizing and extracting themes from the results of the focus groups and interviews. For this analysis, the researcher coded focus group and interview data, linking the raw data to the categories of interest for the study. All of the focus group sessions and some of the interviews were recorded and transcribed. The analysis of themes emerging from focus group and interview data served to identify gaps between the design and implementation of Local Law 196's site-safety program and the needs of immigrant construction workers, specifically day laborers. Elements of the logic model framework were used to evaluate process and outcomes of the site-safety program (Rossi, Lipsey, and Freeman 2004).

This study used a time-series design to explore the impact of Local Law 196 on workplace safety outcomes. The time frame selected for the study included pre-intervention and post-intervention periods. The cutoff date was October 16, 2017, in which at least one element of the new policy went into effect (proctored online training) (NYC DOB, Information Session). A simple OLS regression analysis was used to estimate impacts of Local Law 196's implementation on injury rates in the New York City construction industry before and after the policy change.

Study Limitations and Strengths

Because of the two postponements in the implementation of training requirements under Local Law 196, this study (ended in August 29, 2019) could not assess the new policy's full impact on workplace safety outcomes. Nevertheless, this study achieved objectives related to evaluation of the implementation process by analyzing qualitative data on key aspects of the program, including training content and delivery, and impacts on workers' safety awareness.

The delay in the phasing in of the new policy required that this study emphasized the use of qualitative strategies over quantitative tools, focusing on evaluation of the implementation process rather than estimation of policy impacts. Another limitation of the study related to the data: an increase in number of injuries might reflect an increase in reporting. Lastly, limits to generalization of the study findings also arise from possible self-selection bias in the focus group sample.

Results

Background on Local Law 196: Goals, Assumptions, and Outcomes

New York City's Local Law 196 sets new training requirements for workers and supervisors at jobsites at which the NYC DOB requires construction superintendents, site-safety coordinators, or site-safety managers. The implementation of the law involved three main milestones. The first was a deadline of March 2018 for workers in construction sites to have a minimum of 10 hours of safety training (OSHA-10). The second was a deadline of December 1, 2019, when workers were required to have at least 30 hours of safety training (OSHA-30), and supervisors will need 62 hours of safety training. This milestone was initially set for December 1, 2018, but was postponed twice, first to June 1, 2019, and then to December 1, 2019. The third milestone is September 1, 2020, when workers will need 40 hours of training to be allowed to work at the regulated construction sites. Those 40 hours may include OSHA-30 plus an additional ten hours of training in fall prevention (eight hours) and alcohol abuse prevention (two hours). Decisions to postpone the deadline for the second milestone were made after determining that there was insufficient training capacity to satisfy the new requirements.

Structure and Process

The agency in charge of implementing and enforcing Local Law 196 is the NYC DOB. In September 2018, the NYC DOB created the Construction Safety Compliance (CSC) Unit, which performs inspections for compliance with Local Law 196 in addition to the routine inspections of building code compliance. This new unit checks whether workers and supervisors hold the training identification cards, which are proof of completion of the newly required training. As part of its enforcement function, the CSC issues civil penalties and stop-work orders at noncompliant sites.

A key provision of Local Law 196 created a program to increase opportunities for workers with less-than-equal access to training. To implement this provision, NYC DOB developed a two-pronged approach: (a) the creation of a pool of course providers approved directly by the department to provide site-safety training and issue site-safety identification cards, and (b) a partnership with the NYC Department of Small Business Services (NYC SBS) to certify and oversee training providers for workers with limited access. To achieve its goals under this program, NYC SBS focuses on funding and monitoring training provided by organizations serving day laborers, new entrants (including those recruited through local hiring provisions in affordable housing projects), and minority- or women-owned businesses. The training for less-than-equal access workers is provided to the target population at no cost, with funding from the NYC SBS.

Local Law 196 also provided for the creation of a Site-Safety Training Task Force, which would receive and review recommendations from the public in relation to training and would issue recommendations on training content and delivery to the NYC DOB (City of New York 2017). The task force has 14 members, in addition to the NYC DOB commissioner, who functions as the chairperson. Members of the task force include representatives from the New York City Mayor's Office of Immigration Affairs, the building trades unions, the unionized employer association (Building Trades Employers Association), non-union employer associations, and one representative from a day-laborer organization.

The underlying theory of change of the new policy consists of improving workplace safety outcomes by increasing access to training (especially for vulnerable workers) and by strengthening enforcement. Thus, the administering agencies focused their efforts on providing resources in these two areas, training and enforcement. Program resources and outcomes are discussed below.

Training Capacity

As of June 2019, NYC DOB had approved 72 training providers, including for-profit and nonprofit companies, colleges, and local unions. To build training capacity for less-than-equal access workers, the NYC SBS contracted with day-laborer and other community-based organizations, having provided them with $5 million in funding since mid-2018. For more than three years (including prior to this funding), worker centers in the five boroughs of New York City were receiving funds through a discretionary budget line of the city council to provide safety training to low-wage and immigrant workers. These two sources of public funding enabled worker centers to meet the surge in worker demand for safety training during the past 18 months. For example, the number of workers who received safety training at the Brooklyn-based day-laborer organization Workers' Justice Project (WJP) increased by about 62%, from 691 in 2017 to 1,116 in 2018. From January to August 2019, WJP had already provided safety training to 580 workers. A worker center in Queens, New Immigrant Community Empowerment, provided OSHA-10 and OSHA-30 training for 731 workers from March to December 2018. A worker center in Staten Island, La Colmena, almost doubled the number of workers provided with safety training since 2017, conducting 42 training sessions and training 704 workers from early 2018 to mid-2019. Overall, these three day-labor organizations trained more than 3,100 workers from early 2018 to mid-2019. Accounting for the efforts of all day-labor centers and community-based organizations participating in the program, NYC SBS reported that thousands of workers in the target population received safety training at no cost over the past year (personal communication, NYC SBS official, August 1, 2019).

Monitoring and Enforcement

The NYC DOB expanded its monitoring and enforcement capacity and budget, increasing the number of inspectors and establishing the CSC unit dedicated to safety compliance inspections. As of December 2018, the CSC unit determined that a total of 192 workers did not have OSHA-10 cards at the 167 inspected sites, resulting in the issuing of 482 fines. Overall, 75% of all sites inspected by CSC in 2018 were in compliance (personal communication, NYC DOB official, June 24, 2019).

Program Benefits

According to worker center leaders, a benefit from the program has been the opportunity for their members to become certified to teach OSHA-10 and OSHA-30. In their view, having trainers from similar backgrounds lead the classes has significant advantages over training by individuals who do not share the experiences of vulnerable workers. In addition, the teaching of OSHA-30 has empowered workers to participate in the enforcement of safety regulations at the worksites (personal communication, worker center leader, January 4, 2019).

Remaining Barriers to Successful Implementation

Remaining barriers to access and lack of training capacity drove the day-laborer community to organize protests against the implementation in June 2019 of the second program milestone, which required a minimum of 30 hours of safety training. The milestone was postponed to December 2019, with the possibility of further delay to 2020. The day-laborer community argued that implementation of the second milestone would have resulted in widespread firings of workers who had yet not taken OSHA-30 training (National Day Laborer Organizing Network 2019).

Training Capacity

All stakeholders interviewed under this study agreed that the implementation timelines were not realistic given the limited training capacity and inadequate infrastructure at the program's start. This explained the two postponements of the deadline for its second milestone. The mismatch between the program goals and timelines exerted significant pressure on day-laborer centers' training capacity in particular. The funding provided through a request for proposals process through the NYC SBS began after the first program deadline of March 2018, and it was not until mid-2019 that the New York City Council passed a law authorizing day-labor centers to become certified providers of site-safety training (New York City Council 2019).

Training Content and Delivery Methods

Day-labor advocates believe that the additional eight-hour training in fall prevention, which is part of the ten-hour course required to complete the total of 40 hours of training, is redundant, as this topic is covered by OSHA courses (personal communication, with day-labor center leader, July 8, 2019). Another shortcoming of the additional ten-hour training is that it is planned to be delivered online. Worker center leaders believe that this format is not appropriate for the target population due to literacy issues and lack of experience with the use of computers (personal communications, day-labor center leaders, January 4, 2019; July 8, 2019). Worker center leaders favor the use of popular education methods, as in their experience such methods have been effective in safety trainings for rank-and-file workers (personal communication, day-labor center leader, January 4, 2019). Popular education is a participatory approach to adult education, which promotes critical thinking to empower individuals to overcome societal challenges (Mayo 1997).

Monitoring and Enforcement Challenges

Although NYC DOB has increased budget allocations for site inspections, the agency has limited experience in monitoring compliance with labor standards, including worker safety standards. Consistent with broader organizational goals, NYC DOB increased its staff from 1,551 in fiscal year (FY) 2017 to 1,903 in FY 2019. These additional resources enabled the agency to conduct an increased number of inspections: from 156,508 in FY 2017 to 203,077 in FY 2019 (NYC Mayor's Office of Operations 2019). The agency's unit in charge of monitoring compliance with Local Law 196 focuses on certification requirements (i.e., checking whether workers on construction sites have the required certification cards).

Because the department's primary charge is enforcing building codes focused on public safety, it lacks experience in enforcing labor standards. This has created concerns among immigrant worker advocates, who also indicated that they had minimal input in the policy formation and implementation process. The Site-Safety Training Task Force includes only one representative from a worker center serving the target population. According to day-laborer center leaders, this underrepresentation of the day-laborer community has resulted in missed opportunities to develop collaborative enforcement between their organizations and the NYC DOB (personal communications, day-labor center leaders, January 4, 2019; July 7, 2019).

Limited Outreach

For program administrators, outreach to the target population remains the main challenge, as day laborers and other vulnerable construction workers are hard to reach and lack trust in government agencies. Worker center leaders point out that restrictions in the contract funding from NYC SBS prevent the use of these funds for outreach, drastically limiting the centers' ability to promote access for vulnerable workers (personal communications, day-labor center leaders, January 4, 2019).

Training-Cost Barriers

There is lack of clarity in the language of the new law as to who (employer, worker, or other) is responsible for paying for the newly required training, but de facto, the financial burden and obligation to comply falls on the worker. According to day-labor center leaders, the cost of OSHA-30 courses remains as a barrier for thousands of workers who lack access to the no-cost training offered under the Local Law 196 program (Personal Communications with Day-Labor Center Leaders, January 4, 2019; July 7, 2019; National Day Laborer Organizing Network 2019). An unintended consequence of the new law, as relayed by day-labor advocates, has involved employers who hire certified trainers to issue 30-hour safety training cards by showing up to the sites and speaking to the workers—but only for a few hours. The employer would then deduct the cost of a full 30-hour course (between $180 and $300) from the workers' pay (Personal Communications with Day-Labor Center Leaders, January 4, 2019; July 7, 2019; National Day Laborer Organizing Network 2019). This lack of clarity in the language of the law has also created confusion among workers, who reported cases in which employers who paid for their training were retaining the certification cards, telling workers that they would lose such cards if they went to work for another employer (focus group session, January 25, 2019).

Remaining Language Barriers

Although Spanish is the main language spoken by the program's target population, there is a need to expand access to communities more comfortable in other languages such as Creole, Hindu, Mandarin, Polish, and Russian (personal communication, worker center leader, January 4, 2019).

Findings from Focus Groups

Themes that emerged from the focus group sessions with rank-and-file workers included perceived benefits from safety training offered in Spanish, limitations of the Local Law 196 safety program in enforcement systems and other implementation aspects, and the importance of know-your-rights education, particularly for addressing the specific vulnerabilities of non-union immigrant workers. Participants also identified missing elements for more effective implementation and enforcement, including the need for employer training to promote a safety culture at the workplace and increased emphasis on know-your-rights and leadership education for workers. The vast majority of focus group participants (77 out of 82) had received safety training at no cost at day-labor centers in Brooklyn, Queens, and Staten Island. Of those who received safety training, 36 had completed OSHA-10 training and 41 had completed OSHA-30. Except for two English-speaking workers, all focus group participants were Spanish speakers with limited or no English skills. Women workers comprised 22% (18 out of 82) of total participants.

Benefits of Safety Training

All participants who had received safety training reported increased ability to identify risks related to handling of hazardous materials, electrocution, falling or rotating objects, falls from heights, and in excavation and demolition jobs. They also reported having gained knowledge about proper use of work tools and personal protective equipment. According to at least five of the participants, the new knowledge helped them to avoid near-miss accidents. Workers also reported spillover effects of their safety training, as they shared their knowledge with nontrained and new entrant workers at the job sites.

Participants also reported having gained a sense of empowerment from the safety training, and they relayed instances in which they confronted supervisors about unsafe work conditions, possibly preventing accidents. Workers praised OSHA-30 for providing them with knowledge about their rights and the resources to file complaints. One worker commented, "The training helped me a lot because [now] I take care of myself. … I mean, before I did OSHA-30, I would go to a worksite and do whatever they told me to do. Even if it was at my own risk. But with these trainings, I know what I can do and what I cannot do" (focus group session, January17, 2019).

Implementation and Enforcement Shortcomings

While workers recognized the benefits of safety training, they also pointed out the weaknesses of the overall safety program, particularly as they relate to enforcement systems and the workers' unique vulnerability as non-union and immigrant workers. Most common irregularities that the participants mentioned included employers not providing adequate protective equipment, asking workers not to show up in anticipation of inspections, pushing workers to do unsafe work, and even retaliating against workers who complained. An example of employer retaliation included a focus group participant who said, "the supervisors would always tell us to go on the scaffold without a harness and lifelines. After I learned in the OSHA training that this was wrong, I told them. But they fired me soon afterwards" (focus group session, July 11, 2019). An example of employers neglecting protective measures included an account by a worker, who said, "when I was putting on my protective equipment, a supervisor rushed me and told me 'are you going to take half an hour to get ready to work half a day for me? You know what? Just leave'" (focus group session, January 17, 2019).

Participants said that it is common to work for employers who have no knowledge of safety regulations and that some employers would hold cards that they obtained by taking the online training or had bought in the informal market. A worker in a focus group session said, "sometimes contractors only get the online training for OSHA-30, but they don't [really] know the regulations. … but they get the card. So, when I raised the issue with my boss, he told me 'Don't worry … that does not matter. What matters is that I have my card'" (focus group session, January 25, 2019).

Under the new regulations, the onus is on the worker to undertake the training and hold cards. Thus, workers participating in focus group sessions felt that current enforcement systems exert more pressure on workers (on whether they hold training certification cards) than on employers who should be closely monitored for compliance with safety regulations. A worker said, "contractors don't have Scaffold 4 training. I am required to have it, and I go to work for people that do not have this training. The city should be more forceful [in] requiring contractors to have training certifications" (focus group session, January 25, 2019).

Implementation of the new regulations face additional challenges in the informal sector of the construction industry. Workers in focus groups reported that violations are even more predominant and contractors are even more difficult to track and monitor in the informal sector of the construction industry, than in regulated sites. A focus group participant commented, "[w]e [day laborers] are facing an employer that is more… informal. The regular employer who has an office may pay you little money, but he would give you basic guarantees… your eight hours of work per day, your lunch hour, etc. …The informal employer does not have an office, or an address… He will not give you any guarantees… He tells you 'do you want to work? Work for me today, but don't come tomorrow'" (focus group session, January 17, 2019).

Other concerns about implementation and enforcement issues included the costs of training and the irregularities committed by employers who pay for the workers' training and deduct the cost from paychecks or retain the certification cards if workers leave the job. Participants' comments on these issues included the following: "There were cases in which the employers told the workers, 'I paid for the card. The card is mine. If you leave I keep the card.' Even though they deducted the cost of the training from the paychecks. There has to be a sanction against employers like these" and "Employers should be fined for the practice of retaining safety identification cards" (focus group session, January 25, 2019).

Perceptions About Training Delivery Methods

Focus group participants expressed concerns about the adequacy of the online format planned for delivery of the additional ten-hour modules on fall prevention and alcohol abuse. These are the ten additional hours needed to complete the total of 40 hours of the newly required site-safety training. Participants mentioned factors such as limited literacy, age, and multiple languages spoken by the target population, which would impede access to training delivered online. Comments from a focus group session (January 18, 2019) included these:

- "There are literacy issues that would make online training useless. … Some workers might be able to navigate a website, but they cannot read well."
- "Age is also a problem when it comes to using computers for online training. … There are a lot of older workers for whom technology can be a barrier."
- "[There are] Indigenous populations from Guatemala who don't speak Spanish. They also experience significant barriers for accessing training only offered online."
- "In-person training and group training is much superior to online [delivery methods]."

Perceptions About Outreach

Workers and organizers at day-labor centers point to the limited outreach that the city government implemented to inform workers and employers about the new requirements and to provide orientation about where to get the training. According to focus group participants, the limited outreach created a lot of confusion among employers, who began asking for OSHA-30 cards in March 2018, when only OSHA-10 was required. According to workers and day-labor center organizer reports, many workers were sent back home due to this lack of clarity in the requirementsvand unscrupulous employers took advantage of the situation. According to day-labor organizers, "If workers had known or been well informed, they at least could have told their employers. … [I]n some cases, workers were not being paid for three weeks in a row, and then this happened on top of the nonpayment" (focus group session, January 25, 2019). Day-labor centers were the only source of information about the new requirements for day laborers participating in this study. In focus group sessions, they reported that "there was no outreach [during early implementation phase], … [t]he authorities should have done outreach. If it were not for the worker center, I would not have known. There was no outreach through the social media, or anything" (focus group session, January 25, 2019).

Estimating Policy Impacts with Quantitative Data

Over the past four years, construction employment grew at an annual average rate of 4 percent, adding more than 20,000 jobs from June 2015 to June 2019. Defining a pre-intervention period of eighteen months preceding the policy change, and a post-intervention period of eighteen months following the policy change, means for injury rates were estimated and compared. Average injury rates showed a non-significant increase of 3.1 points from the pre- to the post-intervention period. This increase might be explained by the surge in construction accidents during 2018 (NYC DOB Accident Reports).

Using a simple OLS regression, this study found that increases in construction activity are associated with a 0.96 increase in injury rates (per 100,000 workers), holding all other potential confounders constant. Using this estimated coefficient, predicted values of injury rates were produced and compared with their observed values (Figure 1). Figure 1 shows that increases in construction employment functioned as a relatively efficient predictor of changes in injury rates during the first semester of 2018. In late 2018, actual injury rates become systematically lower than the predicted values (which were based on the pre-implementation period trend).

FIGURE 1

Predicted vs. Observed Injury Rates (per 100,000 workers), Post-Enactment of Local Law 196

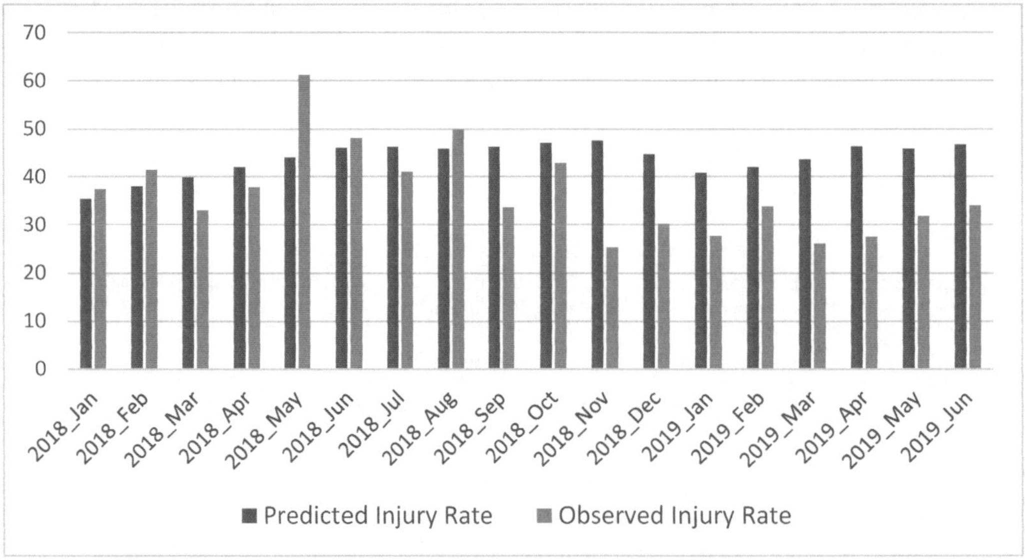

Simple examination of time-series plot of injury rates (per 100,000 workers) before and after the implementation of Local Law 196 can serve to identify effects that might be associated with the newly required site-safety program. Figure 2 shows injury rates for the first six months of the years before and after enactment of Local Law 196. The time series for injuries in 2019 shows relatively more stability than those of previous years and its values are significantly below those of 2018.

FIGURE 2
Injury Rates per Month (per 100,000 workers), First Six Months, 2015–2019

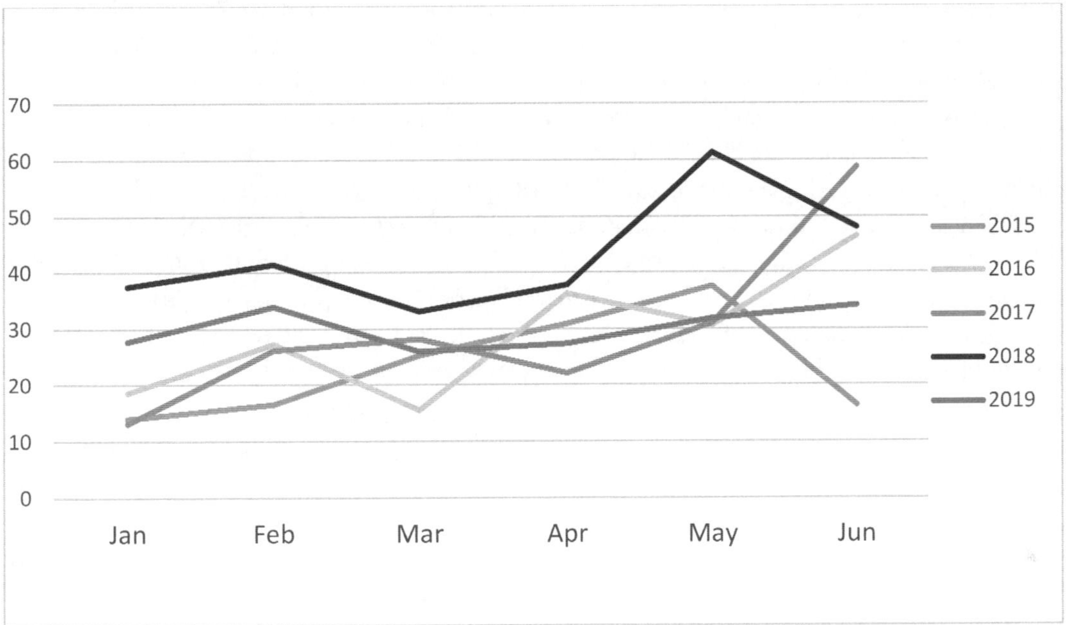

Source: Author's calculations using data from the NYC DOB accident reports and NYS DOL current employment statistics.

The downward trend in injury rates that this study identified is consistent with workplace safety experts' observations, which indicate that fluctuations in construction activity no longer correlate with variations in the number of fatalities in New York City, "as construction is booming yet fatalities are falling" (Obernauer 2019: 6). However, it is important to qualify these results by indicating that the decline in injury rates identified in this study can only be partially attributed to the implementation of Local Law 196. Other factors that might underlie the decline in injury rates include the following: effects of the implementation of mandatory OSHA-10 training in large New York City construction projects since 2007 (Obernauer 2018), increased enforcement efforts by local district attorneys in collaboration with the state's Department of Labor (New York County, District Attorney's Office 2015), and increased training capacity of day-laborer centers resulting from city funding through the Day Laborer Initiative, which was established in 2015 (Nevarez 2015).

Conclusion

Focus group findings revealed the critical importance of safety training to improve safety outcomes of vulnerable and immigrant workers in the construction industry. These findings are consistent with previous research on the benefits of safety training delivered in the language of the target population, as well as on challenges that are specific to immigrant workers, including language barriers, limited literacy, and fear of employer retaliation (Ruttenberg and Lazo 2004). The analysis of qualitative data from focus groups and interviews underscored the limitations of safety training in the absence of other injury prevention mechanisms such as effective enforcement of worksite safety conditions and promotion of a safety culture among construction managers. This finding is also consistent with existing research (Schoenfisch, Hester, and Sinyai 2016).

The qualitative analysis also uncovered a range of areas for improvement in the new safety program, from stronger know-your-rights components of the curriculum to better enforcement and monitoring systems that adequately address the issues that vulnerable workers face. The selection of the NYC DOB as the agency in charge of administering the program was a sensible approach from the perspective that linking compliance to the issuance/renewal of permits would strengthen enforcement. However, NYC DOB's lack of experience in workplace safety and workers' rights necessitates increased collaboration with agencies experienced in monitoring and enforcing labor standards. Training DOB inspectors in occupational safety and health and worker rights, as well as requiring the use of a worksite safety/worker rights checklist in DOB inspections, might strengthen the current enforcement system. Additionally, this study found that NYC DOB's accident classification system does not adequately capture worksite risks that existing safety trainings address. OSHA emphasizes the "fatal four," but the NYC DOB accident classification system does not clearly account for such factors. The available indicators therefore might not be the appropriate metrics to evaluate impacts of the new policy, and so, some modification of the classification system might be needed.

The quantitative analysis in this study identified likely impacts of the new policy on injury rates. The qualitative analysis indicates that for the observed decline in injury rates to be sustained, changes in program design and enforcement approaches might be needed. Such changes include, but may not be limited to, an increased input from day-laborer centers, increased intergovernmental collaboration (i.e., between the Department of Buildings and agencies with experience in enforcement of labor standards), and workplace safety training for building inspectors in charge of monitoring compliance with Local Law 196.

Acknowledgments

This paper was supported by the Center for Construction Research and Training (CPWR) through National Institute for Occupational Safety and Health (NIOSH) cooperative agreement OH009762. Its contents are solely the responsibility of the author and do not necessarily represent the official views of CPWR or NIOSH. The author is grateful to The Workers' Justice Project, New Immigrant Community Empowerment, and La Colmena Staten Island Community Job Center for their assistance in facilitating focus groups of construction day laborers. The author would like to thank the New York City Department of Buildings, the New York City Department of Small Business Services, the Building and Construction Trades Council of Greater New York, and the National Day Laborers' Organizing Network for their insights on the policy formation and implementation. The findings, analysis, conclusions, and recommendations contained herein are those of the author and do not necessarily reflect the views of any of these organizations.

References

Bernhardt, A., DeFilippis, J., Milkman, R., Heckathorn, D., Auer, M., Gonzalez, A.L., Narro, V., Theodore, N., Perelshteyn, J., Polson, D., and Spiller, M. (2009). Broken Laws, Unprotected Workers. Center for Urban Economic Development, University of Illinois at Chicago; National Employment Law Project; UCLA Institute for Research on Labor and Employment, 2009.

Castano, J. (2018). Deadline Arrives for Day Laborers to Get Safety Training. Voices of NY. Feb. 28, 2018. https://bit.ly/3ew3xlu

City of New York, Local Laws of the City of New York for the Year 2017, No. 196. https://on.nyc.gov/3et7O9r

Cornell Worker Institute (2016), Study of Day Labor Hiring Sites in New York City. Interim report. Unpublished.

Creswell, J. (2014). Research Design. Qualitative, Quantitative, and Mixed Methods Approaches. Los Angeles: Sage.

Gonzalez-Rivera, C. (2016). Where Immigrant New Yorkers Go to Work. Data Brief. New York Center for an Urban Future. https://bit.ly/3l0T1Fu

Darragh, A.R., L., Stallones, P., Bigelow, L., and T. Keefe (2004). Effectiveness of the HomeSafe pilot program in reducing injury rates among residential construction workers, 1994-1998. American Journal of Industrial Medicine, 45(2): 210-217.

Dong, X., Entzel, P., Men, Y., Chowdhury, R., and S. Schneider (2004). Effects of safety and health training on work-related injury among construction laborers. Journal of Occupational and Environmental Medicine, 48(9): 914-922.

Dong, X. and Platner, JW, (2004). Occupational Fatalities of Hispanic construction workers from 1992 to 2000. Am J Ind Med, 45(1): 45-54.

Eastern Research Group, Inc. (2014). The Social and Economic Effects of Wage Violations: Estimates for California and New York. Final Report. Prepared for U.S. Department of Labor.

Evanoff, B. and Kaskutas, V. (2013). Construction fall fatalities can be prevented. Occupational Health and Safety, 82(1):34-36.

Guest, G., Namey, E., and K. McKenna (2016). How many focus groups are enough? Building an evidence base for nonprobability sample sizes. Field Methods, 29(1): 3-22.

Jacobson, P. and Wasserman, J. (1997). Tobacco Control Laws, Implementation and Enforcement. Washington, D.C.: RAND.

Johnson, K. and J. Ruppe (2002). A job safety program for construction workers designed to reduce the potential for occupational injury using tool box training sessions and computer-assisted biofeedback stress management techniques. International Journal of Occupational Safety and Ergonomics, 8(3): 321-329.

Koonse, T., Dietz, M., and Bernhardt, A. (2015). Enforcing City Minimum Wage Laws in California: Best Practices and City-State Partnerships. UCLA Center for Labor Research and Education.

Mayo, M. (1997). Imagining Tomorrow: Community adult education for transformation. Leicester: National Institute of Adult Continuing Education.

National Day Laborer Organizing Network, (2019). Day Laborer Site Safety Training Coalition: Proposed Solutions to LL196 Implementation Disaster.

Nevarez, G. (2015). New York City Announces Investment, Expansion of Day Laborer Centers, NBC News Latino, New York. https://nbcnews.to/3mSMBZx

New York Building Congress, 2017-2019 New York City Construction Outlook. https://bit.ly/366NiYu

New York City Council, (2019), Proposed Int. No. 720-B

New York City Mayor's Office of Operations (2019), Mayor's Management Report, Department of Buildings. https://bit.ly/2HYYRJf

New York County, District Attorney's Office (2015). Citywide construction fraud task force convened to investigate misconduct in midst of historic development boom. Press Release. August 5, 2015. https://bit.ly/3k0VEFO

New York County, District Attorney's Office, (2017). Statement by Manhattan District Attorney Cyrus R. Vance Jr., and DOI Commissioner Mark G. Peters on Anniversary of Carlos Moncayo's Death. Press Release.

Obernauer, C. (2019). Deadly Skyline, An Annual Report on Construction Fatalities in New York State. New York Committee for Occupational Safety and Health.

Obernauer, C. (2018). Deadly Skyline, An Annual Report on Construction Fatalities in New York State. New York Committee for Occupational Safety and Health.

Rathod, J. (2016). Danger and Dignity: Immigrant Day Laborers and Occupational Risk. Seton Hall Law Review (46:813)

Rossi, P. Lipsey, M. and Freeman, H. (2004). Evaluation, A Systematic Approach. Seventh Edition. Sage Publications.

Ruttenberg, R., and Lazo, M., (2004). Spanish-Speaking Construction Workers Discuss Their Safety Needs and Experiences. Residential Construction Training Report. Silver Spring: CPWR.

Schoenfisch, A., Hester, L., and Sinyai, C., (2016). Effectiveness of OSHA Outreach Training on Construction Work-Related Injury Rates. Silver Spring: CPWR Small Study No. 15-1-PS.

Sokas, R.K., et al., (2009). An intervention effectiveness study of hazard awareness training in the construction building trades. Public Health Rep, 124 Suppl: 160-8.

Valenzuela, a. et al., (2006). On the Corner: Day Labor in the United States. Center for the Study of Urban Poverty, U of CA Los Angeles.

Williams, Q., et al.. (2010). The impact of a peer-led participatory health and safety training program for Latino day laborers in construction. Journal of Safety Research, 41(3):253-261.

Winter, S. (2006). Implementation. In Peters, G. and Pierre, J., Handbook of Public Policy. Los Angeles: Sage.

IX. 2020 LERA Lifetime Achievement Award

JOEL CUTCHER-GERSHENFELD
Brandeis University

I am deeply honored to receive this award, which has special meaning coming from Dennis Dabney. I have known Dennis since I was a doctoral student doing a case study on a new labor–management partnership at the Budd Company, with Dennis as the management co-lead. It has been a source of great joy and learning to stay connected with his career as he moved to DTE and leading up to his leadership role at Kaiser-Permanente, which is arguably the most important labor relations role in the United States today. As with everything he does, Dennis has brought strong leadership and deep values to LERA. So I begin with a thank you and deep appreciation to Dennis.

A LERA lifetime award requires, of course, reflecting on my life (so far) with LERA, which began, believe it or not, before I was born. The Philadelphia chapter of what was then the Industrial Relations Research Association (IRRA) was formed by my parents and a group of friends at my parents' kitchen table in 1955, the year before I was born. I know—now many of you are doing quick calculations. For the next milestone, you have to fast forward to my being a student in the ILR school and occasionally joining my parents (when I was home) for meetings of the Philadelphia chapter. I joined that chapter in 1978, after graduating from Cornell and working as a managing editor for the Labor Relations Press. By the early 1980s, I had moved to Detroit and was a member of the Detroit chapter while working on staff at the Michigan Quality of Work Life Council. By the middle of the 1980s, I was in graduate school at MIT and a member of the Boston chapter (my local chapter again now). So I am a genuine local chapter mixed-breed.

It was in graduate school that I began what may be an unbroken string of attending and contributing to national meetings, including my first appearance in the proceedings in 1985. This was the 1985 spring meeting (back then there were both December and spring meetings), and my paper had an immodest title for a graduate student, which was "Reconceiving the Web of Labor–Management Relations." Re-reading it now, my call for new, more collaborative institutional relations still seems fresh—which might mean I was forward thinking or that I just hadn't advanced in my thinking!

As a graduate student I was invited to attend the Academy of Management doctoral consortium, which seemed like a great idea, but these doctoral students were not my people. The next year, along with a few other students around the country, we launched the beginning of the current student-led model for doctoral consortia (there had been some prior gatherings but not with the current model). I am delighted to see the doctoral consortium continue as a vibrant forum to this day—representing the future of the field. Over the years, I have helped in the launch of the industry councils and in the renaming of the organization as LERA, which then informed the renaming of the Illinois program and the shift, at the 2009 World Congress in Australia, from IIRA to ILERA.

Sixteen and fourteen are my new favorite numbers. It was sixteen years ago, in 2004, that my father, Walter Gershenfeld, received this same lifetime achievement award—so this may be the first parent–child sequence with the award. We also had another parent–child sequence in LERA, which was fourteen years apart—with my dad serving as president in 1995 and my serving in the role in 2009. My dad was a betting man, and I am sure he would figure out something to do with the numbers sixteen and fourteen. The same year of my presidency, 2009, was also the year my mother, Gladys Gershenfeld, received the LERA

Practitioner Award, so family is very much in my mind at this time. I only wish they could be here—and you know they would then be telling everyone about this afterward.

In reflecting on family at this time, please allow me to do an unpaid promotional announcement for the Gladys and Walter Gershenfeld Diversity Fund, supporting under-represented voices for increased participation in LERA meetings and activities. This is a refocusing of the publications fund set up with my brothers, Neil and Alan, in honor of our parents. Mom, as many of you know, was among the first female arbitrators in the field and was dedicated to mentoring women entering the profession. She also helped to rewrite key National Academy of Arbitrators documents in gender-neutral language. Dad's doctoral dissertation, in the early 1960s, was one of the first to document what we now call institutional racism. He was studying employment relations in the city of Lancaster, Pennsylvania, which at the time had a Black unemployment rate of 12.4%—more than four times the White unemployment rate. The question was, even in a prosperous community, why did Black unemployment persist? Interestingly, there was no breakout of unemployment data by race at the time, so my dad calculated this 12.4% Black unemployment rate based on a door-to-door survey, working with a team of Black interviewers, going to every third house in all of the Black neighborhoods. Both Mom and Dad would appreciate this fund, which complements LERA's newly established Diversity and Inclusion Committee.

I would like to thank Tom Kochan, Robert McKersie, Michael Piore, Lisa Lynch, and other mentors of mine in graduate school; and colleagues from MSU, MIT, Illinois, Australia, and now Brandeis, with its amazing commitment to social justice. I share this award first with my wife, Susan; and then with our sons, Gabriel and Aaron; my brothers, Neil and Alan, and their families; and the many friends and family—all of whom are integral to my lifetime achievement.

In conclusion, it is traditional, upon receiving this award, to note that it doesn't mean my life's work is complete. For me, it is not just about looking back but also about looking ahead. In that context, current developments—both the pandemic and the broader challenges in society—point to the need for more agile and effective institutions in society. That is the work now and in the years and decades to come. It is with this work in mind that I accept this award. I thank Dennis and LERA for this great honor.

X. Using Online Job Vacancy Data to Study Labor Market Dynamics

Mismatch in Online Job Searches

MARTHA E. GIMBEL
Schmidt Futures

TARA M. SINCLAIR
George Washington University and
Indeed Hiring Lab

Introduction

Public debate keeps returning to the issue of whether or not there are structural problems in the labor market in terms of a mismatch between the background, skills, and/or interests of job seekers as compared to the needs perceived by employers. The "skills gap" or "talent shortage" conversation often relies on anecdotes because it can be hard to collect data at a sufficiently detailed level to appropriately quantify mismatch. Previous research has provided measures based on connecting data from a variety of different sources with varying levels of detail. Online labor market data provides the potential for new insights based on a single source of rich data on both vacancies and job seekers.

The mismatch index is designed to measure the level of mismatch, or dissimilarity, in the economy. It compares the number of job seekers in a job category, based on their employment history, to the number of vacancies in the same category. Mismatch can arise because there are too few or too many job seekers in a particular category relative to the number of job opportunities. Importantly, our measure of mismatch is relative to the overall availability of job seekers and vacancies. Thus, we are focused here on the mismatch across categories rather than movements in the aggregate job seeker to vacancy ratio which might be affected by changes in the use of online job search platforms in general and/or the market share of a particular platform.

We produce monthly mismatch measures for the United States, a set of English-speaking countries, and select US sectors from January of 2014 through June of 2019. Our main finding is that mismatch has declined as the economy has improved. This decline has been driven primarily by a return of jobs to bring the distribution of jobs more in line with the distribution of job seekers.

Our analysis is closely related to Şahin et al. (2011, 2014) and Lazear and Spletzer (2012a, 2012b) who also quantify the level of mismatch in the economy. They use publicly available survey data from the Bureau of Labor Statistics (BLS) and measure mismatch based on industry categories. They also use vacancy data from the Conference Board's Help Wanted Online (HWOL) index to construct mismatch measures for a set of occupation categories. Other research, such as Burke et al. (2019), uses job postings data aggregated by Burning Glass Technologies for vacancy information. Marinescu and Rathelot (2018) use data from job board CareerBuilder.com to estimate the role of geographic mismatch and find that it plays a minor role in explaining aggregate unemployment.

There has also been substantial research on mismatch outside the United States and particularly in the United Kingdom. Turrell et al. (2018) use data from Reed, an online recruiter in the United Kingdom, to estimate mismatch by occupation and geography in the United Kingdom. They find that regional mismatch rather than occupational mismatch affects UK productivity. Patterson et al. (2016) and Smith (2012) use data from the UK government employment agency JobCentre Plus to construct estimates of mismatch with the Patterson et al. finding that occupational mismatch is an important contributor to weak productivity growth in the United Kingdom and the Smith finding that occupational mismatch has had a substantial impact on UK unemployment rates.

Şahin et al. (2014) focus on measuring "mismatch unemployment"—i.e., the share of unemployment due to sectoral mismatch. For their occupation-level analysis they report results using 22 of the 23 major (two-digit) SOC groups and 36 of 96 minor (three-digit) SOC groups. In the working paper version, Şahin et al. (2011) use the same mismatch formula we use here for a benchmark measure with no heterogeneity across markets. They consider all 17 industries where publicly available vacancy data are available.[1] They conclude that mismatch explains up to one third of the increase in the unemployment rate during the Great Recession.

Lazear and Spletzer (2012a, 2012b) used a measure of mismatch as part of a broader set of indicators on the recent performance of the US labor market. In terms of mismatch, they focused on their finding that mismatch rose in the recession and then declined afterward, suggesting a cyclical rather than structural pattern.

In this paper, we present a set of mismatch indexes that we compare across English-speaking countries (the United States, the United Kingdom, Australia, Canada Ireland, New Zealand, and Singapore). Similar to Lazear and Spletzer, we are particularly interested in what the patterns in our mismatch measures over time tell us about how different types of mismatch are related to changes in economic conditions. With our unique dataset, we can focus on a range of different levels of disaggregation to create different measures of mismatch in terms of geography, sector, and job seeker characteristics.

For example, we include all active online job seekers, both employed and unemployed, in our benchmark series, where we identify a job seeker as someone who updated their résumé on the job search website within that month. Including employed job seekers has been challenging in previous analyses due to limited data availability on people searching while employed.[2] There is debate about how similar employed and unemployed job seekers are and what impact differences might have on economic outcomes. On the one hand, Ahn and Hamilton (2019) argue that the unemployed differ in terms of relevant unobservables for job finding that vary over time, and Longhi and Taylor (2014), using UK data, find that the unemployed and employed are quite different and that the differences vary over the business cycle. On the other hand, Kroft et al. (2016) find that "shifts in observable characteristics of the unemployed do not go very far in accounting for the rise in long-term unemployment." Most related to our analysis, Şahin et al. (2014) see little difference when adding in employed job seekers based on time use surveys into their measure of mismatch.

In addition to mismatch, we also produce measures of vacancy dissimilarity over time as well as job seeker dissimilarity over time. Comparing the distribution of job opportunities today to what was available in the past and doing the same for job seekers gives us a measure of how much the labor market has shifted over time from both the labor supply and labor demand dimensions. This is particularly important given one of our key findings for the United States is that mismatch is declining somewhat over our sample period. At the same time, we find substantial change in the distribution of both vacancies and job seekers over this period, so the slightly declining mismatch suggests that jobs and job seekers are becoming more similar to each other as the economy has improved. We then show that the decline in mismatch is mostly driven by changes on the job posting side, suggesting that missing jobs from the recession have been returning in the recovery in a way that makes the vacancy distribution look more like the job seeker distribution.

In the following sections, we describe our data and mismatch methodology, and then we report our benchmark measure of overall online labor market mismatch for the United States. We find that mismatch has slightly declined as the labor market has tightened, while the distribution of jobs has changed substantially. The changes in the distribution of jobs and résumés have overall drawn job seekers and employers closer together over the sample. We also provide results for a set of sectors as well as cross-country comparisons. We then conclude with a discussion of future work.

Data

The analysis is focused primarily on the United States, but we also include analysis for the United Kingdom, Ireland, Australia, New Zealand, Singapore, and Canada. Our main data source is online job postings and job seekers from Indeed, the largest job site in the world based on unique visitors according to comScore, an independent analytics firm.[3] For comparison, we also use publicly available data from the Current Population Survey (CPS) and the Job Openings and Labor Turnover Survey (JOLTS).[4] We focus on seasonally unadjusted data from all sources. Our measure of mismatch will be in shares of totals, which should net out any common seasonal patterns, and will leave only job category seasonal patterns, which we are interested in examining.

Our measure of job openings will either be from JOLTS by industry, where we focus on the 12 industries and where we can match with data available from the BLS on the industry of the unemployed, or from job postings aggregated by Indeed from across the Internet.[5] The Indeed postings number for each month is the average daily postings visible on Indeed in that category for that month. We also considered job postings visible on the last business day of the month to line up with the definition from JOLTS but found that it was typically similar to average daily postings and that using the average daily posting number smoothed out any single-day effects. We also compared all visible postings to only those from employer websites (excluding job boards whose visibility on Indeed has varied over time) and found the results to be similar.

It is important to note that we are not restricted to advertisers on Indeed. Instead, Indeed collects job postings anywhere on the Internet and de-duplicates them as part of their business. Indeed is a generalist site in the sense that they focus on providing "all jobs" not a niche market.

Our measure of job seekers will either be the (experienced) unemployed, classified by the industry of their last job (from the CPS provided by the BLS), or active job seekers (both employed and experienced unemployed) on Indeed, classified by their most recent job title on their résumé uploaded on Indeed. Indeed has 64.7 million résumés from the United States as of June 2019. We are focusing on the subset that were active accounts during our sample from 2014 through June 2019, where "active" is defined as having last updated their résumé on Indeed in that month.[6] We aggregate to the monthly frequency, but we could look at daily or even intra-day based on the Indeed data. Higher frequency is interesting when looking at the job seeker data.[7]

For robustness, we also use an alternative measure of job seekers based on clicks on job postings. A job seeker can click on a posting only if a job is available, and the click may not indicate the job seeker is qualified, only that they are interested in the role. We then classify the job seeker based on the titles of jobs they click on and compare the distribution of clicks to the distribution of job postings. This analysis allows us to use all job seekers on Indeed rather than being limited to account holders.

Job seekers are not just the unemployed.[8] In fact, it appears that the majority of job seekers on Indeed are employed based on reported employment status by account holders as well as reported in internal surveys. This is consistent with the finding by Faberman et al. (2017) that employed job seeking is "pervasive." We identify labor market status in the Indeed data based on information reported by the user. Users opt-in to being counted as employed by checking a box indicating that they are currently employed at one of the positions listed on their résumé. There is likely measurement error as some employed workers may not select the box and others may try to hide that they are unemployed by selecting the box or by not updating that information if they leave their employer but continue searching for a job on Indeed. Therefore, we do not report separate results for employed and unemployed job seekers but only combined results for all job seekers. We include only the "experienced unemployed" in our résumé data because we are only using résumés that have previous employment recorded. This is consistent with the CPS data where an industry is only available for people who were previously employed. For our clicks analysis, however, the clicks can come from any job seeker and we do not observe their current employment status.

In the online labor market data, we have much finer job type groupings than what is available in the data used in previous research: for our benchmark measure, we include 6,068 normalized title pairs per month in our analysis as compared to the 9 to 36 categories used by Lazear and Spletzer (2012b) and Şahin et al. (2014). For example, "registered nurse" is a normalized title that contains registered nurse, RN, RN staff nurse, registered nurse (RN), registered nurse–RN, registered nurse traveler, etc. "Economist" is a normalized title that contains economist, health economist, principal economist, chief economist, associate economist, lead economist, and so on. The 6,068 titles were determined as the superset of English normalized titles across the countries in this study: the United Kingdom, the United States, Canada, Australia, Ireland, New Zealand, and Singapore. For some titles, the counts for both résumés and postings are zero in most or all months for one or more countries, which does not meaningfully affect our analysis. We also estimate a version excluding low observation categories with no meaningful impact on the estimates. We organize our analysis around job titles for a number of reasons: (1) titles are relatively easy to standardize across résumés and job postings and across countries, (2) titles capture skills more consistently that what is reported by job seekers in résumés, (3) employment background provides a blend of interest and skills to better connect with where a job seeker will likely go than just a narrow classification of job seekers by skills alone, and (4) titles allow us to get quite granular as compared to industries or occupations.

Methodology

The mismatch measure is the Duncan and Duncan (1955) dissimilarity index. With this measure, we assume that only the job seekers can change occupation, whereas job vacancies are fixed in their category.[9] The Duncan and Duncan measure is

$$\frac{1}{2}\sum_i \left|\frac{S_i}{S} - \frac{V_i}{V}\right|, \qquad (1)$$

where S_i are the job seekers in category i, S is the total number of job seekers, V_i is the number of vacancies in category i, and V is the total number of vacancies.

This is the same measure used by Lazear and Spletzer (2012a, 2012b) and Şahin et al. (2011, before incorporating a matching function). This index can be interpreted as the proportion of job seekers who would need to be moved to make the job seeker to posting ratio the same for all job categories, where a job category in our analysis will either be industry or normalized job title. Other measures of mismatch, notably Şahin et al. (2014), are reported as a fraction of hires lost per period due to job seeker misallocation. Thus, our index will likely be much higher in magnitude as a share of job seekers as compared to a share of monthly hires.

Benchmark Results

For our measure of mismatch based on online job search, we start in January 2014 and report through June 2019.[10] One of the benefits of using the online data is more timely arrival of updated information. As soon as the first week of each month, we could update our measures rather than waiting for JOLTS vacancy data, which arrives over a month later and then is revised further in the following months when later surveys come in. JOLTS vacancies are further revised annually all the way back to the beginning of the series in December 2000 to incorporate updates to the Current Employment Statistics employment estimates. Seasonally adjusted data are also revised with updated seasonal factors, but we are using seasonally unadjusted data throughout.

Figure 1 presents our online labor market mismatch estimate along with industry mismatch based on unemployment from the CPS and vacancies from the JOLTS following a similar methodology to that used by Lazear and Spletzer (2012a, 2012b). Our measure is higher in level, as would be expected given that we are moving from 12 industry categories to over 6,000 job title categories. In terms of time pattern, however, they are broadly similar, although our measure is substantially smoother.

FIGURE 1
Comparing Mismatch Measures, January 2014–June 2019

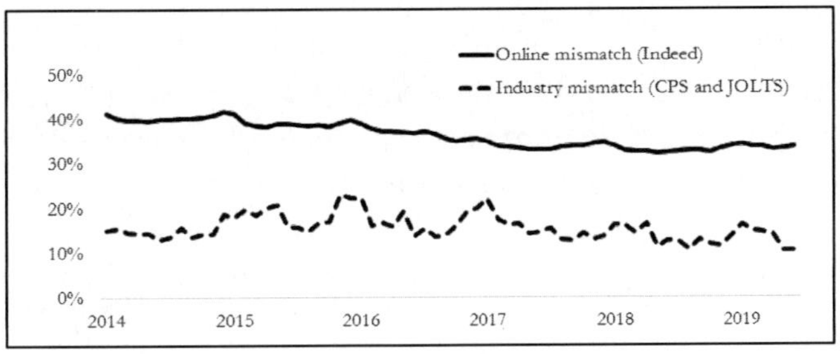

Source: Indeed data, BLS, and authors' calculations.

Lazear and Spletzer find much more mismatch by occupation than by industry, which is consistent with what we find for our online labor market mismatch at the normalized job title level. Job titles are much more similar to occupation than to industry. We would also expect that there would be more mismatch at lower levels of aggregation.[11]

We have explored a number of different groupings and our results are consistent with what is expected: grouping the job titles into broader categories (Indeed's proprietary categories) results in a lower level of mismatch overall (as seen below in Figure 2) but a similar pattern of slight decline over our time frame. Limiting the analysis to only titles with large numbers of postings and résumés (e.g., the top 700) gives very similar results in both level and slope, which is consistent with how mismatch is measured because it is driven by large categories. It is also similar in terms of smoothness, which suggests it is not the large number of titles that is driving the smoothness of online mismatch as compared to industry mismatch based on publicly available data.

FIGURE 2
Comparing Online Mismatch Measures
[Job seeker and posting shares grouped by titles (6,068) or categories (57)]

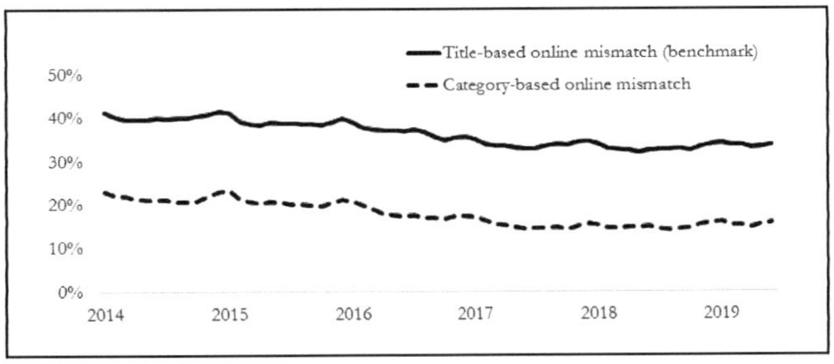

The smoothness of online mismatch may be due to the consistency of the data since the online data source is a common labor market with as much as possible the same definitions applied to both groups. It does not appear to be sensitive to changes in aggregation level, the particular dissimilarity metric used, or changes in our definition of an active job seeker.[12] There appear to be seasonal movements in the distribution of unemployed job seekers in the CPS that are different from the JOLTS job openings numbers, which results in seasonal fluctuations appearing in the industry mismatch series. One interpretation of the smoothness in the online mismatch series is that employed job seekers may have fewer seasonal differences from openings as compared to the unemployed, but further analysis beyond the scope of this paper would be needed to confirm that interpretation.

Despite the smoothness, we do see clear seasonality in mismatch. This might be expected because we do not use seasonally adjusted data, but it is interesting that the seasonal patterns are sufficiently different in job postings versus job seeker behavior that we see clear rises and falls each year in our mismatch measure.

At least three concerns arise from our use of the latest job on job seekers' résumés in order to classify them. The first is that job seekers may be aware of the changing landscape of job opportunities and they may be looking for roles different from their current or most recent job title. The second is a concern about the way the résumé data are stored that may be affecting our results. Per the terms of Indeed's user agreement, only the latest résumé a job seeker has uploaded is kept. That means we lose some of the earlier job seekers in our sample since we count an active job seeker based on the month the résumé was last updated.[13] Third, using résumé data means we limit the sample to job seekers who have uploaded a résumé on Indeed, but many people use the website without uploading a résumé. To address these concerns, we consider an alternative measure of job seeker distribution based on the job titles job seekers click on. This allows us to focus on the jobs a job seeker is looking for rather than their experience. The job seekers may not always be qualified for the roles they look at, so the clicks-based measure is more about interest, whereas the résumé title captures work experience. Another caveat of this measure is that job seekers cannot click on a job if they are not shown the role, so the clicks are affected by both job posting availability and the Indeed search algorithm.

Despite the caveats and substantial differences between our two different job seeker measures, the mismatch series created by using the same job posting shares as before and measuring job seeker shares in the two different ways are surprisingly similar. As shown in Figure 3, clicks mismatch is lower than résumé mismatch early in the sample, but, by 2017, the two measures are very similar. Both show some decline over time, but it is more muted for the clicks measure. This leads us to emphasize "not increasing" rather than "clearly declining" in interpreting our US results.

FIGURE 3
Mismatch with Different Job Seeker Measures
(Click shares captures interest for next role vs. experience in résumé)

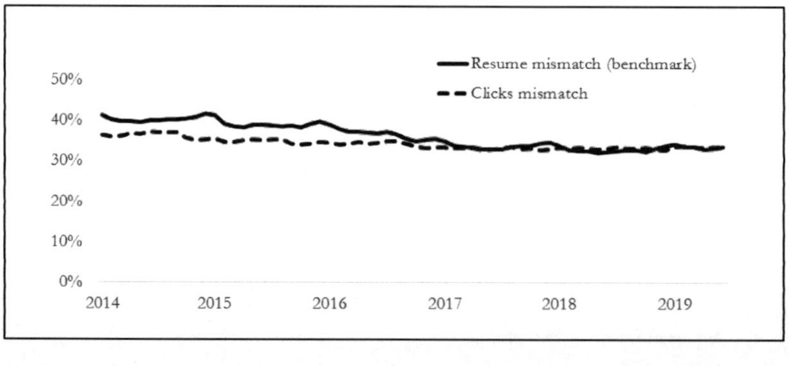

Looking into the normalized titles that are the largest contributors to mismatch, presented in Table 1, a few features stand out. First, these titles are large categories. This is important to keep in mind for the dissimilarity measure we use—it is based on differences between the shares in the postings and the résumés, so even a large percentage difference in a small category would not result in a large move in overall mismatch. The top ten where the résumé share exceeds the posting share contributes 10.9% of mismatch, and the top ten where the posting share exceeds the résumé share contributes 10.2% of mismatch. The top contributors to mismatch are also notably persistent, with some seasonal patterns. For example, comparing this list to the list for December 2018, we get slightly different ordering but remarkably similar titles with the exception of "seasonal associate" appearing prominently in the December list for posting share exceeding résumé share. Comparing June 2019 mismatch contributors with June 2016 results in substantial overlap, with over 50% of the same titles showing up on both the 2019 and 2016 lists.

TABLE 1
Top Contributors to Online Mismatch
[Comparing job seeker résumés and job postings in June 2019 (Indeed data)]

Rank	Résumé share > Posting share	Posting share > Résumé share
1	Customer service representative	Retail sales associate
2	Cashier	Shift manager
3	Customer service associate/cashier	Registered nurse
4	Server	Restaurant manager
5	Receptionist	Babysitter/nanny
6	Warehouse worker	Assistant manager
7	Laborer	Shift leader
8	Forklift operator	Store manager
9	Manager	Restaurant staff
10	Nursing assistant	General manager

Changing Job Postings and Changing Résumés

Mismatch could be flat to declining for two reasons: either little is changing underneath or job seekers and jobs opportunities are seeing their distribution across titles change in similar ways over the past several years. To examine this, we used the same dissimilarity index but applied it to jobs and résumés separately over time to see how different jobs and résumés are today from what they were in 2014. Thus, for each time period *t*, from January 2014 through June of 2019, we constructed the following dissimilarity metric:

$$\frac{1}{2}\sum_i \left|\frac{V_{i,t}}{V_t} - \frac{V_{i,2014m1}}{V_{2014m1}}\right|. \tag{2}$$

We find that the jobs mix has changed substantially over the past few years. The job seeker mix has also changed, although not as dramatically. Overall, as we show below, it is the change of job postings toward job seekers that has brought about the small decline in mismatch over the sample.

First, looking at the distribution of job postings over time: Figure 4 shows that there has been a substantial change in the distribution across titles in job postings over recent years. Comparing January 2019 with January 2014 (comparing January to January to exclude potential seasonal differences), 25.8% of job postings in 2019 would need to change in order to have the same distribution as five years before.[14]

FIGURE 4
Changing Mix of Job Postings over Time
(Evolution of US job postings mix over time)

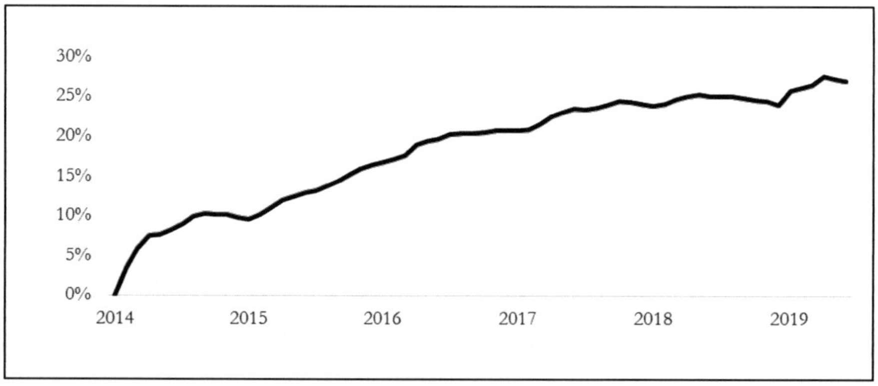

Résumés have changed less over the sample than job postings have. Again comparing January 2019 with January 2014, résumés are 15.0% different than they were five years before (Figure 5). One data note: because of the nature of Indeed's data, where only the latest résumé a job seeker has uploaded is kept, résumés today are less comparable with résumés five years ago than job postings over the same time period.

FIGURE 5
Changing Mix of Online Résumés over Time
(Dissimilarity of job titles in résumés compared to January 2014)

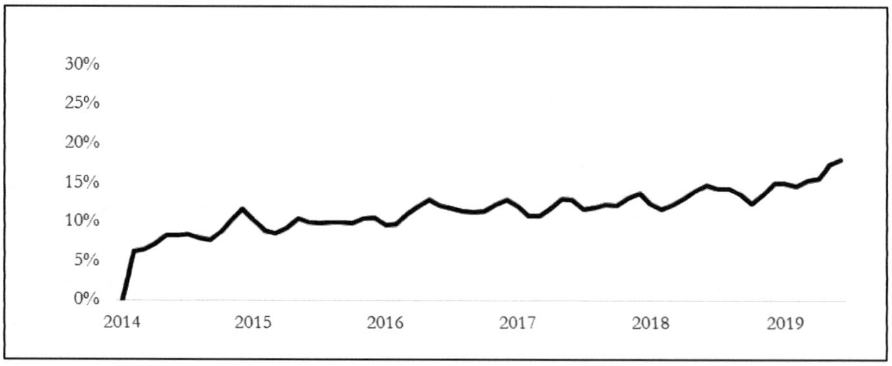

In order to explore the role of the changes in postings and résumés and their contribution to mismatch, we constructed counterfactual mismatch measures where we held the labor supply (résumés) or the labor demand (postings) distribution constant at the shares of the beginning of the sample (January 2014). Figure 6 shows that mismatch would have been a bit higher in 2019 if the résumé distribution had not changed, but it is much more dramatic when we hold the postings distribution constant: in that case, mismatch would have risen rather than declined over the sample.

FIGURE 6
Analysis Holding One Side of Mismatch Constant
(Mismatch holding one side at January 2014 shares)

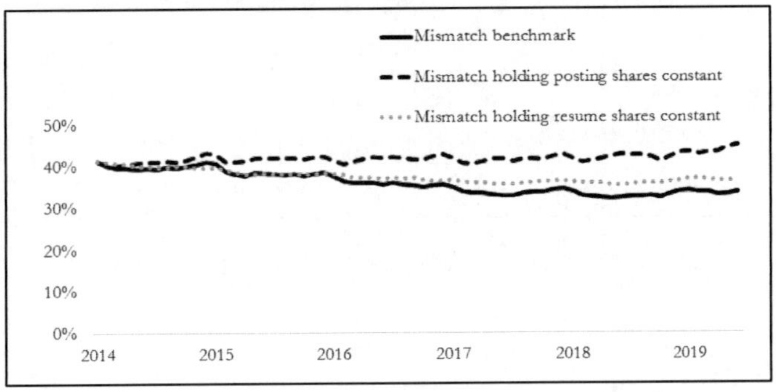

Sector Analysis

We can also explore the question of how well matched the job seekers are within the sector, which we might think of as sort of intensive margin mismatch.[15] For the within-sector mismatch we return to our dissimilarity measure and calculate mismatch based on résumés of job seekers currently or most recently employed in that sector and job postings in that sector. Each sector is defined by a set of normalized titles that can clearly be mapped to that sector. Our three sectors are tech (550 titles), healthcare (289 titles), and finance (571 titles). In June 2019, healthcare was the largest sector, with approximately 14% of all US job postings. Finance had less than 2% and tech had almost 6% of all US job postings.

In Figure 7, we show that for most of the sample, healthcare shows greater mismatch than our benchmark overall results for the United States, and tech and finance are both below. Interestingly, at the end of the sample, healthcare mismatch declines and tech mismatch rises to converge close to the overall national level of mismatch. Finance, however, stays flat and well below the national level.

FIGURE 7
Mismatch for Tech, Healthcare, and Finance Sectors
(Mismatch within different sectors)

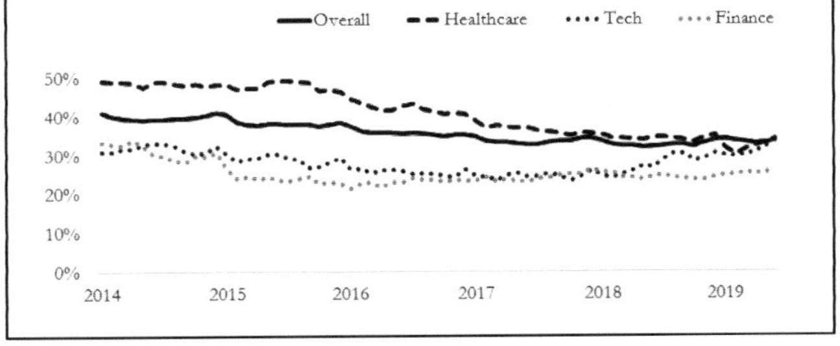

Cross-Country Comparisons

For the same set of 6,068 normalized titles (selected as the superset of normalized titles across the countries), we construct comparable mismatch measures, again monthly from 2014 through June 2019 (Figure 8). The countries have slightly different levels and seasonal patterns, but perhaps the most interesting pattern is the trends: all seven of the countries studied: the United States, the United Kingdom, Canada, Australia, Ireland, New Zealand, and Singapore. Canada and the United States have very similar levels and patterns, with Canada just slightly below the United States throughout the sample. Australia is at almost the same level of mismatch at the end of the sample as in 2014. The similarity of the United States, the United Kingdom, and Canada is consistent with other labor market indicators for these countries over this time period.[16]

FIGURE 8
Within-Country Mismatch Comparisons
(Overall mismatch in online job search by country)

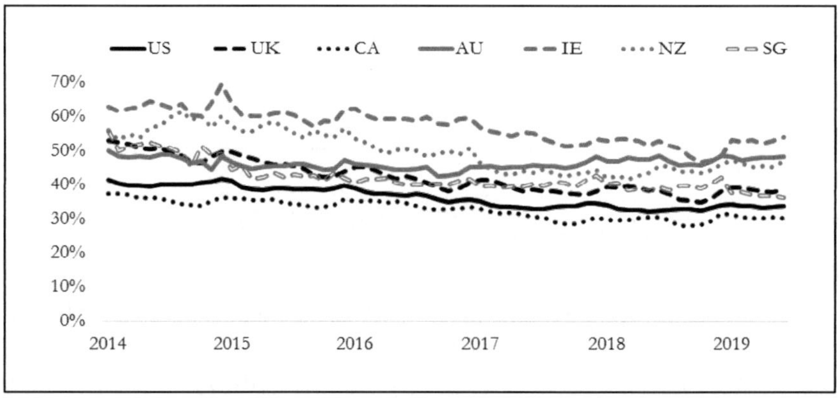

We also constructed the dissimilarity index for job postings over time for each of the countries in our dataset and report the comparison of the results in Figure 9. We see that all the countries have seen a substantial change in the distribution of their mix of job postings between 2014 and 2019, ranging from Australia's 21.7% change to Ireland's 32.3% change (comparing January to January to avoid seasonal differences).

FIGURE 9
Postings Shares Changes over Time for Seven Countries
(Evolution of job postings mix by country)

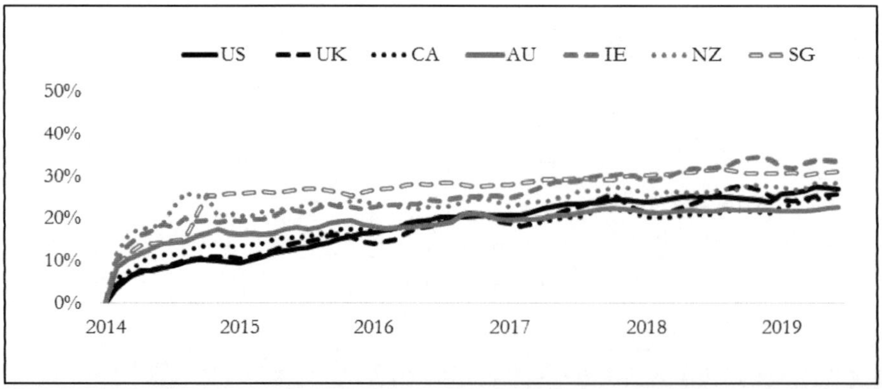

Source: Indeed data.

Conclusion

This paper shows that even though the distribution of job vacancies has changed substantially since 2014, we see a robust trend of slight decline in mismatch between the distribution of online job vacancies as compared to the distribution of online job seekers over the past several years for the United States and across a range of English-speaking countries. The decline in mismatch appears to be driven by the change in the distribution of jobs toward the distribution of job seekers. One interpretation is that jobs came back that were a better fit for job seekers as the global economy continued to improve over the past several years.

This analysis opens up several directions for future work. In particular, this analysis, consistent with Lazear and Spletzer (2012b), suggests there is a cyclical component to mismatch, which means if we knew the trend or natural rate of mismatch, we could potentially use mismatch as an additional indicator of slack. With our estimates only available for a recovery period, we have little business cycle variation to estimate what is trend and what is cycle, but we expect there to be more information along these lines as we update the series over time.

Furthermore, modeling and weighting for potential career changers may provide additional insights. Although we consider a variety of different aggregation levels with robust results, for each set of categories, our analysis is binary: same category or not same category. One concern about grouping job seekers into categories is that job seekers may not stay in the same category and that skills may be transferable across categories and/or job seekers may develop new skills over time that might lead them to change categories. This may be particularly true of the finer categories we use at the normalized job title level. Furthermore, people may have the skills for jobs but be uninterested in doing them (interest mismatch as compared to skills mismatch). Hobijn (2012) combined data from the CPS, JOLTS, and state-level job vacancy surveys and found that the "majority of job openings in all industries and occupations are filled with persons who previously did not work in the same industry or occupation." Sinclair (2014) and Flowers (2018) have both examined the behavior of job seekers using Indeed to search for jobs in categories other than their most recent employment and find substantial amount of searching across even very broad categories. They also each document that specialization and pay are both positively related to retention by job type. This analysis suggests we may want to weight by some measure of skills and/or interest overlap for our mismatch index. In that case, we may be able to think about the distance between normalized job titles and estimate a smaller amount of mismatch in "adjacent" job titles by occupation grouping. A related approach was used by Şahin et al. to allow their unemployed job seekers to search in a new industry/occupation, but they find that the "bulk of unemployed workers keep searching in their previous employment sector" (2014: 3559), so their estimate of mismatch unemployment is little affected. We can also rank order the normalized titles by estimated average salary to construct a weighted variant of the dissimilarity index called the Earth Mover's Distance (Rubner et al. 2000; for an application to the labor market, see Rim 2018) or use a measure of occupational distance such as Robinson (2018).

Finally, we can produce estimates of mismatch for more types of job seekers and more regions and countries. In preliminary work, we have estimated mismatch for all the US states and found a decline in mismatch across all states over our time frame to suggest that the national decline is broad based rather than driven by a subset of states. In future work, further analysis of the state-level patterns may provide additional insights. It may be interesting to zoom in not just on narrower geographies and sectors but also on mismatch by other features of the job seeker. For example, we can look at employment status, long-term versus short-term unemployed,[17] and age categories. Indeed also has data for over 60 countries with broadly similar data collection and structure, so we would like to build indexes that are comparable across countries, although we will have to address how to get consistent job titles across languages.

Acknowledgments

Adhi Rajaprabhakaran provided excellent research assistance and Indeed's Data Insights Platform team provided essential data support. Both authors of this paper had an employment relationship with Indeed when involved in the research.

Endnotes

[1] The 17 industries used by Şahin et al. are arts, construction, mining, accommodations, retail, professional business services, real estate, wholesale, other, transportation and utilities, manufacturing (nondurables), education, health, government, manufacturing (durables), finance, and information. The 12 industries we use in our analysis are construction, durable goods manufacturing, nondurable goods manufacturing, wholesale and retail trade, transportation and utilities, information, financial activities, professional and business services, education and health services, leisure and hospitality, other services, and government. Lazear and Spletzer use 12 industries but differ from ours by including mining but grouping together durable and nondurable goods manufacturing. We exclude mining due to different definitions applied to vacancies and job seekers in the publicly available data. Results are little changed between the different choices of Lazear and Spletzer, Şahin et al., and our analysis.

[2] Şahin et al. (2014) did provide an estimate of their measure including on-the-job search. They used the American Time Use Survey to identify employed job seekers. This survey likely underestimates the number of employed job seekers as discussed in Faberman et al. (2017).

[3] Globally, Indeed has 250 million unique visitors per month (Google Analytics, Unique Visitors, September 2018) and is the #1 job site worldwide according to comScore total visits (March 2018). Indeed has 55.4 million unique visitors per month in the United States (comScore, November 2018), which makes Indeed the #1 ranked job site by unique visitors in the United States. Furthermore, in July 2018, comScore estimated that 75% of US online job seekers search for jobs on Indeed (per month).

[4] The job openings data are from the September 10, 2019, release of JOLTS. The unemployed by industry data are from the CPS. The data are not seasonally adjusted, and using the 12 industries available from both CPS and JOLTS: construction, durable goods manufacturing, nondurable goods manufacturing, wholesale and retail trade, transportation and utilities, information, financial activities, professional and business services, education and health services, leisure and hospitality, other services, and government. Note that we exclude mining due to different definitions between JOLTS and CPS (although including it does not give noticeably different results).

[5] Şahin et al. (2011, 2014) and Lazear and Spletzer (2012a, 2012b) also each produce measures of occupational mismatch using Help Wanted Online Index (HWOL) data as their measure of vacancies for a subset of standard occupation categories (since only industry groupings are available from JOLTS). The HWOL data by occupation are not publicly available and thus we focus on the industry mismatch as our comparison. Canon et al. (2013) provide a review of mismatch indexes using HWOL job vacancy data.

[6] It is possible to use Indeed for job searching without opening an account or uploading a résumé, but our main sample is limited to those with accounts and résumés. Indeed saves only the latest version of résumés, so we count each résumé only one time based on the latest update date because the last job title from the résumé is key to our analysis. We recognize this might cause a bias in the analysis if there is a systematic pattern in who updates résumés frequently and/or who was a job seeker on Indeed early in our sample and again later in our sample. We also estimated a version with the latest résumé attached to all accounts, but activity was determined by the date that the résumé was created. This could also cause a bias because a job seeker could have been in one role in 2014 and searching for a different role, gotten that role, updated their résumé, and searched again in 2016. Interestingly, however, the results were nearly identical in the two models, so there does not appear to be much bias from the updating of the résumés.

[7] There are interesting daily and weekly patterns in the job search data—but less so for job postings data. See https://indeedhi.re/2U55hso for discussion of daily patterns in the data.

[8] We are looking only at active job seekers, so they are either employed or unemployed; there is no "out of the labor force" group in our analysis.

[9] The Duncan and Duncan measure has come under criticism when applied to occupational gender segregation (Watts 1992, 1994, 1998). An alternative measure, the IP index of Karmel and MacLachlan (1988) is the preferred measure in that literature. In the gender segregation case, however, both men and women could change occupations; whereas, in our analysis, we assume only the job seeker can change occupations.

[10] The data from Indeed are only available consistently over time starting in January 2014, and analysis for this version started in June 2019. For discussion of our initial results, see https://bit.ly/3eAujcy.

[11] According to Şahin et al., "Every statement about the role of mismatch should be qualified with respect to the degree of sectoral disaggregation used" (2014: 3538). Comparing across different aggregation approaches (occupation versus industry, for example) and/or across different data sets can also shift the level of mismatch. We are focused less on the level of mismatch and more on the pattern in mismatch over time.

[12] We change the measure of the job seeker below to interest based on clicks and, in additional unreported robustness checks, we also used different dating conventions for identifying an active job seeker with little impact on the results. We also considered an alternative measure of dissimilarity, the Kullback–Leibler (KL) divergence measure (using Bayesian Dirichlet priors; see the recent survey by Yang 2018, for more details on the KL divergence measure) and find broadly similar results.

[13] We also estimated mismatch identifying job seekers using their latest résumé and the date they first uploaded it to Indeed as an alternative and found extremely similar results for mismatch suggesting the updating is not causing much bias.

[14] We also considered our alternative dissimilarity measure, KL divergence. The results are consistent across the two measures, with January 2018 compared to January 2014 having a KL statistic of 0.23 and a similar trend over time.

[15] See these two blog posts for further discussion of the healthcare and tech results: https://bit.ly/2Ibyy22 and https://bit.ly/3l6IzMC.

[16] See also https://bit.ly/2U0VPq3. For more analysis of the Canadian and Australian data, see the following blog posts: https://bit.ly/3k53vlT and https://bit.ly/2U0W4RZ.

[17] Wiczer (2015) argues that occupation-specific shocks are important for understanding the pattern of unemployment duration over the business cycle.

References

Abraham, Katharine G. 2015. "Is skill mismatch impeding US economic recovery?" ILR Review Vol. 68, no. 2, pp. 291-313.

Ahn, Hie Joo, and James D. Hamilton. 2019. "Heterogeneity and unemployment dynamics." Journal of Business & Economic Statistics, pp. 1-26.

Burke, Mary A., Alicia Sasser Modestino, Shahriar Sadighi, Rachel Sederberg, and Bledi Taska. 2019. "No Longer Qualified? Changes in the Supply and Demand for Skills within Occupations." https://bit.ly/2I8Zi3o

Cappelli, Peter H. 2015. "Skill gaps, skill shortages, and skill mismatches: Evidence and arguments for the United States." ILR Review Vol. 68, no. 2, pp. 251-290.

Canon, Maria, Chen, Mingyu and Marifian, Elise, 2013. "Labor mismatch in the Great Recession: a review of indexes using recent U.S. data." Federal Reserve Bank of St. Louis Review, Vol. 95 (May/June), pp. 237-272.

Duncan, Otis Dudley, and Beverly Duncan. 1955. "A methodological analysis of segregation indexes." American Sociological Review, Vol. 20, no. 2, pp. 210-217.

Faberman, Jason, and Marianna Kudlyak. 2016. "What does online job search tell us about the labor market?" FRB Chicago Economic Perspectives Vol. 40, no. 1.

Faberman, R. Jason, Andreas I. Mueller, Ayşegül Şahin, and Giorgio Topa. 2017. "Job search behavior among the employed and non-employed." National Bureau of Economic Research Working Paper No. 23731.

Flowers, Andrew. 2018. "The Top Jobs People Don't Leave." https://bit.ly/3k1kUvz

Galaasen, Sigurd Mølster and Andreas Kostøl. 2018. "Mismatch and the Consequence of Job Loss" Presentation at the Symposium for Nonlinear Dynamics and Econometrics.

Hobijn, Bart. 2012. "The Industry-Occupation Mix of U.S. Job Openings and Hires." Federal Reserve Bank of San Francisco Working Paper No. 2012-09.

Karmel, Tom, and Maureen MacLachlan. 1988. "Occupational sex segregation—increasing or decreasing?" Economic Record Vol. 64, no. 3, pp. 187-195.

Kroft, Kory, Fabian Lange, Matthew J. Notowidigdo, and Lawrence F. Katz. 2016. "Long-term unemployment and the Great Recession: the role of composition, duration dependence, and nonparticipation." Journal of Labor Economics Vol. 34, no. S1, pp. S7-S54.

Kuhn, Peter, and Hani Mansour. 2014. "Is Internet job search still ineffective?." The Economic Journal, Vol. 124, no. 581, pp. 1213-1233.

Lazear, Edward P., and James R. Spletzer. 2012a "The United States labor market: Status quo or a new normal?" National Bureau of Economic Research Working Paper No. 18386.

Lazear, Edward P., and James R. Spletzer. 2012b. "The United States labor market: status quo or a new normal?" In Proceedings, Economic Policy Symposium, Jackson Hole, Federal Reserve Bank of Kansas City, pp. 405-451.

Longhi, Simonetta, and Mark Taylor. 2014. "Employed and unemployed job seekers and the business cycle." Oxford Bulletin of Economics and Statistics, Vol. 76, no. 4, pp. 463-483.

Marinescu, Ioana, and Roland Rathelot. 2018. "Mismatch unemployment and the geography of job search." American Economic Journal: Macroeconomics, Vol. 10, no. 3, pp. 42-70.

Patterson, Christina, Ayşegül Şahin, Giorgio Topa, and Giovanni L. Violante. 2016. "Working hard in the wrong place: A mismatch-based explanation to the UK productivity puzzle." European Economic Review, Vol. 84, pp. 42-56.

Rim, Nayoung. 2018. "The Effect of Title IX on Gender Disparity in Graduate Education." AccessLex Institute Research Paper No. 18-05.

Robinson, Chris. 2018. "Occupational Mobility, Occupation Distance, and Specific Human Capital." Journal of Human Resources, Vol. 53, no. 2, pp. 513-551.

Rubner, Yossi, Carlo Tomasi, and Leonidas J. Guibas. 2000. "The earth mover's distance as a metric for image retrieval." International Journal of Computer Vision Vol. 40, no. 2, pp. 99-121.

Şahin, Ayşegül, Joseph Song, Giorgio Topa, and Giovanni L. Violante. 2011. "Measuring mismatch in the US labor market." https://nyfed.org/3k6bMWG

Şahin, Ayşegül, Joseph Song, Giorgio Topa, and Giovanni L. Violante. 2014. "Mismatch unemployment." American Economic Review, Vol. 104, no. 11, pp 3529-64.

Sinclair, Tara M. 2014. "What Job Seekers Want: Occupation Satisfaction & Desirability Report." https://indeedhi.re/3k3SiSv

Smith, Jennifer C. 2012. "Unemployment and Mismatch in the UK." In Bank of England/Institute of Macroeconomics conference on Unemployment, productivity and potential output: The aftermath of the crisis.

Turrell, Arthur, Bradley Speigner, Jyldyz Djumalieva, David Copple, and James Thurgood. 2018. "Using job vacancies to understand the effects of labour market mismatch on UK output and productivity." Bank of England Working Paper No. 737.

Watts, Martin. J. 1992. "How should occupational sex segregation be measured?" Work, Employment and Society, Vol. 6, no. 3, pp. 475-487.

Watts, Martin J. 1994. "A critique of marginal matching." Work, Employment and Society, Vol. 8, no. 3, pp. 421-431.

Watts, Martin. J. 1998. "Occupational gender segregation: Index measurement and econometric modeling." Demography Vol. 35, no. 4, pp. 489-496.

Wiczer, David. 2015. "Long-Term Unemployment: Attached and Mismatched?" Federal Reserve Bank of St. Louis Working Paper No. 2015-042.

Yang, Jangho. 2018. "Information theoretic approaches in economics." Journal of Economic Surveys, Vol. 32, no. 3, pp. 940-960.

XI. LERA Annual Reports

LERA Executive Board Meeting Minutes
Sunday, January 5, 2020, 10:15 a.m. PT
Manchester Grand Hyatt, Promenade AB, San Diego, CA

Call to order—The meeting was called to order at 10:33 a.m. by Kris Rondeau, President. Present at the meeting were officers Kris Rondeau (President), Dennis Dabney (President Elect), Adrienne Eaton (President Elect Elect), Ryan Lamare (Secretary-Treasurer), Bill Canak (NCAC Chair), and Ariel Avgar (Editor-in-Chief). Board Members in attendance were Matthew Bodah, John Budd (RVP Mid), Bill Dirksen, Cyndi Furseth (Membership Committee Chair), Brad Markell, Dan Marschall, Jim Pruitt (Development Committee Co-Chair), David Weil, and Jeff Wheeler. Aaron Sojourner, Committee Chair for the LERA@ASSA Program Committee, was also present. LERA staff attending the meeting were Emily Smith and Bernadette Tiemann.

Approval of the Minutes—The motion to approve the minutes passed unanimously from the board meeting in June 2018 and the General Membership Meeting in June 2018; Brad Markell motioned and Matt Bodah seconded.

Reports

LERA@ASSA 2019 Program Committee Report—Aaron Sojourner reported. LERA is one of the founding organizations with the ASSA and as such we have a disproportionately large number of sessions allocated to us at the ASSA meeting, and this is an important advantage to maintain open dialog with other economists and policy-makers that meet at this conference. Currently, the ASSA uses a four-year average of median attendance numbers to re-allocate sessions, so attendance is important to maintaining our session allocation at its current level. The program committee is working to improve marketing of our sessions to achieve good attendance, and is also focused on planning excellent sessions, with an array of diverse speakers and research represented. We received $4,500 in sponsorship to offset the cost of the reception and 59 members joined or renewed in conjunction with presenting on this program (45 of these people joined as new members). Bill Canak reminded staff to be sure to send information on new members joining from this program so that they can be encouraged to join their local chapters as well.

Finance Report—Ryan Lamare reported that 2018 has been LERA's strongest year financially in recent times. Our net income was $62,000, which is our third year in row of reporting a net gain. Gains from the last three years have been directed towards refilling the organization's reserves that had been drawn down in prior years.

Dues are higher than projected for 2018; we are up over 100 members from last year, although library subscribers are down. Meeting income is also higher than projected for 2018, partly due to another conference held in conjunction with our own sponsored by the LERA Health Care Industry Council, and other special events that were sponsored, but held at cost. The special events were a positive experience for members but did not add to expenses passed through to meeting registrants, as they were sponsored events. There was a cost savings from *Perspectives on Work*; this was because production management was brought in-house and handled by Bernadette Tiemann on staff.

The 2019 budget has been built with conservative estimates on where we expect our income and expenses to be, but we do expect to be in the black for 2019 as well. We may be seeing a degree of leveling off for 2019 in terms of joins and renewals, but hope to maintain and slightly grow in the future. Ryan Lamare presented the 2017 findings of the auditors for discussion, which had normal findings. The auditors have suggested that we adopt a more traditional accounting methods of reporting our restricted assets so that the balance sheet has a better visualization of what assets are available to the organization. The board will review the various restricted funds at the next board meeting in June, as well as the allied agency agreement with the University of Illinois. Ryan Lamare called for a closed executive session.

After the full meeting re-convened, it was disclosed that the following were approved unanimously by the board: promotion and pay increase for Bernadette Tiemann, pay increase for Emily Smith, and an approval of general salary program as appropriate from the University.

Membership Committee Report—Cyndi Furseth reported that we have had an increase in our member numbers of about 100 people this year, in addition to the increase last year of about 100 members. We have seen a slow-down in the last quarter of 2018 so we may or may not see this type of increase next year. If board members have ideas about membership promotions for 2019 and beyond, we are open to ideas. Apprentices were added as a new category last year, and we currently have 7 apprentice members. Jeff Wheeler suggests that bringing in students as speakers to help bring them into this professional atmosphere. Recruiting young members to help direct the social media of the organization is also helpful. Some chapters allocate 20% of regular member dues to support young members membership and registration to the conference meeting. Having students on committees would be quite helpful, including the membership committee. Let's attempt to get every committee involved with a student. One of the positive membership outcomes in 2019 is the linkage between LRAN and the LERA. LRAN will be holding their conference in Cleveland the day before the LERA 71st Annual Meeting, and Erin Johannson was also just elected to the LERA board. It might be beneficial to develop a strategy to approach LRAN members and help them join the LERA.

Nominating Committee Report—Kris Rondeau reports that the nominating committee has been formulated and requests input from the board of directors. This group of individuals is designed to be independent of the board. David Weil suggests that we invite Aaron Sojourner to be invited to also act on the nominating committee 2019. We will also invite a member or officer of the student LERA chapter at Rutgers University to act on this committee.

In the last election, the following new board members and officers were elected: Adrienne Eaton, (President Elect 2019-20); and board members: Kati Griffith (Academic, 2019-2022); Jake Rosenfeld (Academic, 2019-2022); Heather Boushey (Neutral, 2019-2022); Erin Johnsson (Labor, 2019-2022), Daniel Altchek (Management, 2019-2022), and Robert Chiaravalli (RVP Mid, 2019-2022).

LERA 71st Annual Meeting Program Committee Report—Dennis Dabney and Harry Katz are serving as the Program Committee Co-Chairs for the LERA 71st Annual Meeting in Cleveland, Ohio. Dennis Dabney reported that the committee process of proposal selection was efficient this year, and the Cleveland program has been scheduled and is ready to go. The contract has been signed for the Portland meeting in 2020, with a food and beverage commitment of $30,000 which is manageable and a room block of about 650 room nights, with an 80% attrition rate to meet in order to receive all our concessions in the contract. There are no planned increases in registration pricing for 2019, as the LERA annual meetings have been making ends meet for the last three years. The Labor Research Action Network (LRAN) is planning to conduct their annual meeting on Wednesday the day before the LERA Annual Meeting, also in Cleveland, to coordinate with us.

Editorial Committee Report—Ariel Avgar, LERA editor-in-chief since 2015, reported that the LERA 2018 Research Volume and the Perspective on Work Vol. 22 were both released in late November/early December 2018. Each year, the LERA research volume features scholarly research, examining one topic in-depth, and Perspectives on Work focuses on short, magazine-like articles that directly appeal to our practitioner members. This volume of Perspectives on Work successfully made strides towards being more inclusive in our range of contributing authors, which was a concern brought up at our last board meeting in Baltimore.

The editorial committee continues to have two LERA research volumes in production at any given point. The 2019 volume is currently being produced on the topic: "Employment and Disability: Issues, Innovations, and Opportunities". It's edited by Susanne Bruyere, Lisa Yang, and Hock Tan, all from Cornell University. The 2020 volume is on "Reimagining the Governance of Work and Employment" and will be edited by Dionne Pohler from the University of Toronto. In June, the editorial committee will solicit proposals for the 2021 volume. The committee is engaged in reviewing the model for soliciting proposals and is open to suggestions on improvement/innovation.

The LERA editorial committee also works in partnership with the ILR Review to create a "LERA Best Papers" special section in the ILR Review ... which may at some point become a special volume. Papers presented at a LERA meeting are eligible to participate in this competition. From the 2016 LERA meetings, the competition received 11 papers, of which 6 were reviewed, and 2 made it into the ILR Review. Another paper is likely to be approved from the 2017 conference. We have discussed investing in a scholarly LERA journal of our own, but so far this arrangement is proving to be a productive outlet for scholarly publication for our academic members.

National Chapter Advisory Council Report—William Canak, LERA NCAC Chair, reports that LERA chapters are healthy and engaged, with about 40 chapters. We now have a LERA Virtual Chapter, running through Penn State University, which has been conducting meetings using Zoom technology. Tom Kochan spoke at their inaugural meeting with 29 virtual attendees. We will continue to see how they innovate and supply LERA with new resources. The Maine LERA Chapter has established their own website with good features/functionality. The TERRA chapter has been engaged to collaborate with the Tennessee Labor-Management Foundation (TNLMF), with the goal of elevating their annual conference programming. With TERRA's help, Steven Greenhouse presented at the TNLMF Sept. conference, as did Jack Clarke (NAA). In return for assistance with programming, the TNLMF will help with marketing, logistics, and local arrangements at the TERRA conference.

The NCAC will host a chapter representatives meeting and a chapter administration workshop at the LERA 71st Annual Meeting in Cleveland. The RVP's have been functioning well (John Budd, Jim Pruitt, Michele Hoyman). Robert Chiaravalli has just been elected to succeed John Budd to represent the mid-region chapters. He has been a very engaged chapter member and past board member of LERA.

The NCAC proposes that the LERA invite, each year, all the local LERA chapter presidents to attend the annual LERA meeting, and reimburse them for registration following the event. The chapter could send another officer in lieu of the president should that be expedient for the chapter. The NCAC feels that this could be a good investment in the LERA chapters and would keep open the pathways of communication between LERA chapters and the national organization. Matthew Bodah moved the motion to the floor and it was seconded by Kris Rondeau. The motion passed unanimously with no exceptions:

> *Motion:* To encourage interaction between local LERA chapters and LERA national members, facilitate exchange of communication at all levels of the organization, expose LERA Chapter Presidents to resources available through national LERA, and provide a conduit of information from the LERA National Chapter Advisory Council (NCAC) to LERA chapter officers/members, LERA will invite LERA Chapter Presidents to attend the LERA Annual Meeting and to attend the LERA Chapter Representatives Meeting, held at the annual conference. LERA Chapter officers may vote to send a designated chapter

elected officer other than the current Chapter President. After the Chapter President or designated, elected chapter officer attends the LERA annual meeting, including the LERA Chapter Representatives Meeting, LERA's conference registration fees will be refunded after the conference. Only one officer per chapter is eligible for this reimbursement.

Development Committee Report—Jim Pruitt, Development Committee Co-Chair, reported that Marlene Heyser, also Co-Chair of the committee, was unable to attend the board meeting as she is currently recovering in the hospital from a serious accident and head injury. We are hoping for a full recovery for Marlene, and many of our members are checking in on her frequently.

The LERA Annual Fund Drive in 2018 was budgeted at $20,000 but brought in $26,000. We received sustaining sponsorship funds in the amount of $45,000 from both UAW/Ford Motor Co. and Coalition of Kaiser Permanente Unions. These sustaining sponsorships have meant a great deal to the health of this organization and we deeply appreciate this support of our mission.

Our organizational members increased from 46 to 59. The committee continues to encourage organizational membership from multiple organizations including AFSCME, Longshore, UFCW, Teamsters, Steelworkers, Unite Here, among others. The work with trying to get more organizations represented within LERA continues with the development committee for 2019.

Our major university contributors included Cornell University and Rutgers University, and Illinois is considering moving into this category for 2019. Other California university programs are also exploring organizational membership with the LERA.

LERA kicked off a new Major Gift Campaign in late 2018 as well, and we will continue to pursue major gifts from individual contributors as well as final gifts and estate gifts. Jim Pruitt made a personal contribution to the major gift campaign for 2018 and invited all other board members and officers to join him in making personal contributions as well, many of whom have done so.

Next meeting will take place on Friday, June 14, 2019 in Cleveland, Ohio, in conjunction with the LERA 71st Annual Meeting.

Adjournment—The meeting was adjourned at 1:12 p.m. by President Kris Rondeau.

LERA Executive Board Meeting Minutes
Friday, June 12, 2019, 6 p.m. ET
Via Videoconference

Call to order—The meeting was called to order at 6:04 p.m. Dennis Dabney, President. Present at the meeting were John Amman, Ezio Borchini, William Canak, Robert Chiaravalli, Paul Clark, Joel Cutcher-Gershenfeld, Dennis Dabney, Bill Dirksen, Virginia Doellgast, Adrienne Eaton, Shannon Gleeson, Kate Griffith, Cyndi Furseth, Erin Johansson, Ryan Lamare, David Lewin, Wilma Liebman, Brad Markell, Jim Pruitt, Emily Smith, Maite Tapia, Bernadette Tiemann, Andrew Weaver, and David Weil.

Approval of the Minutes—The motion to approve the minutes passed unanimously from the board meeting in January 2020. Seconded by William Canak and approved unanimously.

Reports

Executive Committee Report—Dennis Dabney reports that the Diversity and Inclusion Committee recommends that LERA issue a statement on anti-racism. The committee has appointed Dennis Dabney to compose a statement and it will be emailed to the board for review and approval.

Membership Committee Report—Our membership decreased due to COVID-19 by 150 members between February and April 2020. At that point, LERA began promoting our COVID-19 webinars. Between those 14 webinars, the all-day *Virtual ADR from A to Z* training which was a first-of-its-kind joint membership promotion between LERA chapters and national LERA, and the virtual LERA 72nd Annual Meeting, we have been delighted to see 200 members join in the last 90 days. As a result, we are now at a new high, at least for the last few decades or so for which we have reliable data. The membership committee believes that the new videoconference webinar medium is important to continue developing as a membership benefit and membership promotion. David Lewin suggests that a few webinars on police unions and police reform might be a current topic to pursue. Brad Markell would encourage involving the AFL-CIO into this process. The membership would also like to conduct quarterly committee meetings, discuss founding a membership program, and some form of on-boarding for new members. Joel Cutcher-Gershenfeld suggests that there is a public sector interest section that could be tapped to organize this session or series of sessions. Kate Griffith suggests that it would be good to collect the current research on the topic into a collective repository.

Financial Report—LERA Secretary-Treasurer Andrew Weaver reports that 2019 was a strong year. The organization increased its scale, even before taking into account. The big challenge as we all know came this year with COVID-19 and the hit to membership and the large change in our meeting revenue. However, we have now recruited more members than were lost, and we have taken in enough meeting registration to make ends meet in 2020. We feel this is a tremendous accomplishment despite the challenge that 2020 has brought. In general, things are manageable and we will have a positive year. This is further strengthened by two fairly substantial donations to our quasi-endowment, one at the end of 2019 and one at the beginning of 2020.

Program Committee Report—Wilma Liebman reported further discussion at today's committee meeting on the theme for the LERA 73rd Annual Meeting; the theme continues to be discussed by the committee even now. The chair challenged everyone on the committee, and on the executive board to organize a session to be submitted in November 2020 for review, and also asked a number of people on the committee with Michigan ties (including Bill Dirksen, Bob Chiaravalli, Maite Tapia, and Lynn Rhinehart) to form a cohort to address a range of regional topics and other matters pertaining to the location. Some of the themes discussed will be essential workers and their working conditions (warehouse workers, fast-food workers, low-wage workers), police workers, etc., and "is this a transformative moment?" Over the last several months issues related to workers and labor have been in the news more than they have been for years. Other critical topics include safety and health, unemployment, depression. Additionally, Black Lives Matter or the NAACP will be involved. Panels should include both academics and practitioners. The committee would like to feature something on the tech industry (AI, facial recognition, etc.). The Call for Proposals will be distributed shortly and will emphasize these topic areas and others, and the committee is open to proposals and suggestions from the board. Bill Dirksen contributed that UAW/auto industry should be involved at the Detroit meeting. Bob Chiaravalli added that a tour of the Detroit Institute of Arts Museum, specifically for the Diego Rivera's masterpiece Four walls which features 27 paintings on Detroit industry. A plenary session with two or three labor historians would be fascinating, including Nelson Lichtenstein and others. Virginia Doellgast encourages international and comparative subjects, specifically addressing the pandemic and responses to it. Bill Canak also discussed Joel Cutcher-Gershenfeld reprising an event held in Chicago a number of years ago about changes in work. Brad Markell suggested a session discussing NAFTA. Adrienne Eaton and the Diversity and Inclusion Committee recommend that the program committee include a statement about diversity and have drafted something for the board to consider later in this meeting, and that racial equality and justice should be called-out themes in the Call for Proposals. Brad Markell recommended that everyone read an open letter penned by Bill Spriggs.

Editorial Committee Report—Ryan Lamare reports on the LERA/ILR Review Best Papers submission process has resulted in positive outcomes. There are a few papers that have been published based on LERA presentations; two this year were awarded that were presented in 2018. One paper from the 2019 meetings is still in the review process. The 2020 research volume edited by Dionne Pohler is on schedule; it is almost out the door. The 2021 Research Volume is being edited by Tobias Schulze-Clevin, and one-third of the chapters are received, with the goal of the remainder to be received by the rest of the summer. The proposals and ideas for the 2022 volume was discussed in committee, and one was based on the idea of racial injustice, the protest activities we see around racial injustice and how work can be viewed within this larger story. Maite Tapia, MSU, is interested in being part of this project and contributing with Tamara Lee at Rutgers University.

Development Committee Report—Jim Pruitt reports that the organization brought in $36,000 in contributions. The motion to establish a quasi-endowment fund passed unanimously, seconded by Adrienne Eaton, and is as follows:

> "The LERA Executive Board at its June 12, 2020 meeting hereby establishes a quasi-endowment fund for purpose of providing ongoing stability and income for operations, preserving gift assets without subjecting them to substantial risk, and providing additional real growth through new gifts. The Fund will support on-going operational expenses and board designated projects consistent with LERA's mission. Contributions will be reported each year to the Development and Contributions Committee and to the Executive Board. The list of board designated projects will be reviewed every 5-10 years. Asset allocation will be reviewed periodically by the Secretary-Treasurer and the fund will be managed to provide ongoing income and preserve principal. Spending from the fund should not exceed 2-5% annually depending on growth, general economic conditions, and organizational needs. The spending policy should be reviewed every 5-10 years. A 20% administrative fee will apply to contributions. LERA will use the online contributions site at the LERA website, member renewals forms, and member renewal programs on the web to inform possible contributors. All contributions to the will be recorded in the LERA financial reports and thank you letters will be sent to contributors."

Jim Pruitt encouraged fellow board members, officers, and committee chairs to give a current contribution to the organization if they have the means to do so.

Industry Councils-Interest Sections Coordinating Committee Report—Joel Cutcher-Gershenfeld, IC/IS Coordinating Committee Chair, reports that the COVID-19 webinars were very well-attended, but better attended at the beginning than at the end, which indicates some saturation for the medium that should be considered when planning any future webinar activity. The various industry councils and interest sections involved are receiving an infusion of new leadership. If you review the rest of the included report you will see people who have a passive interest in a various areas. Now that we have this C3 tool, it will be much easier for section and council leaders to reach out to these individuals to organize them. Dennis Dabney remarked that because this has generated new membership, we encourage the sections and councils to continue with the webinars. Paul Clark discussed conducting quarterly webinars to bring industry councils and interest sections together throughout the year. John Amman would like to pursue creating an arts and entertainment industry council, and potentially combining it with sports.

Nominating Committee Report—Joel Cutcher-Gershenfeld, Nominating Committee Chair, reports that the slate is confidential to the board. What you see in the slate a group of individuals who represent the potential candidates for the 2020 LERA election. Diversity in a number of areas was addressed and we consider this a balanced slate. David Weil seconded this slate the slate was approved unanimously. Increasingly we have presidents who are activists in the role, and if a president wishes to appoint a small committee to assist them, that can certainly be done, but that group should work exclusively with the president as to not create additional demand on staff. Joel mentioned that the Walter and Gladys Gershenfeld fund was recently re-purposed to support diversity and inclusion within LERA, and to defray the costs of participation to under-represented groups in ways that will not create more work for staff.

National Chapter Advisory Council Report—William Canak, NCAC Chair, reports that the state of the chapters is good with an exceptional number of chapters are receiving awards this year. Presentation of chapter awards will take place at the chapter representatives meeting. Chapters have been good at paying their dues to LERA in 2019; a few are in arrears but are have been communicated with. No chapters became inactive this year. When we first established Regional Vice Presidents in 2016 to represent chapter interests on the board, it was decided to stagger their elections, elect one each year. This year we will elect our RVP for the West region. The NCAC encourages establishment of more student chapters, and additional student members in all LERA chapters. LERA members are required to join their local chapter, and we certainly hope that all national board members, officers, and committee chairs are complying with this. Virginia Doellgast asked for more information about what a student chapter does, and the response was that they have speakers, events, occasionally conduct joint events with other chapters and in most senses function similarly to other LERA chapters.

Regional Vice Presidents Report—Jim Pruitt (RVP West) and Robert Chiaravalli (RVP Mid) reported. Since last year, Robert has met twice over the phone with the Milwaukee chapter on how they can increase attendance, and meetings with DC LERA as well. A video conference will be conducted for remaining chapters in the coming months. Jim Pruitt reports that NorCal LERA has renewed is activities under Renee Mayne and Sam Kim's leadership; NorCal's LERA's group of leaders is terrific. COVID-19 pushed off their inaugural event and they will try to re-create the event when safe. The other chapters in the region are doing well.

Diversity and Inclusion Report—Adrienne Eaton brought additional recommendations to the board for review.

The first recommendation gives members the ability to report their own demographic data so that we can begin reviewing this data as an organization. "LERA give members the opportunity to report their gender, race, ethnicity and disability status on membership forms."

Dennis Dabney moved to accept a motion for item one and Jim Pruitt seconded the motion and it passed unanimously.

The second recommendation encourages more diversity within our session panels and at our meetings generally.

A) LERA will include a note in future Calls for Proposals encouraging organizers to consider diversity in organizing panels and include resources for finding diverse presenters (Econ and Political Science have websites for this purpose.) Draft language for CFP: "Ideas for stimulating and creative proposals related to engaging various, sometimes differing, stakeholders involved in employment relations with the end goal of enriching investors, managers, employees, policy makers, and unions are invited. LERA's program format offers fresh opportunities for thought leaders from all aspects of employment and industrial relations to create connections. LERA's mission bridges research, practice and policy, as well as stakeholders. Incorporating these three aspects is strongly encouraged, as is a balance of gender, racial and generational diversity among the proposed participants."

B) LERA will add the following resources and similar ones that members identify to the website: https://econspeakerdiversity.shinyapps.io/EconSpeakerDiversity/ and https://womenalsoknowstuff.com/. Further, LERA will consider building a similar resource for our community.

C) The LERA submission form will ask panel organizers to comment on the diversity of their panels in their submission (e.g. including gender, race, generational, academic/practitioner) DRAFT LANGUAGE FOR CFP: Organizers are required to provide a session abstract and description, identify participants (confirmed or invited), comment on diversity on the panel and provide full contact information at the time of submission for all session participants.

D) LERA Program Chairs and Committees will be encouraged to take into account diversity when a) deciding which panels to include in the program, b) making recommendations for changes to panels and c) selecting featured speakers.

Item 2, with sub-sections A, B, C, and D above passed with no changes, though there was one opposed.

Lastly, the committee recommends that LERA appoints people to leadership and confer awards who exhibit good behavior, and awardees should be advised that they will need to inform LERA if they have a finding of sexual harassment. Concerns were about those who might be under current investigation and subsequently found innocent and/or people who have reformed and are later in life.

A) "LERA will encourage nominations for awards and elected positions and appointments for committee chairs and members only be made to persons with a record of ethical and professional behavior."

B) "LERA will advise nominees or appointees for awards and leadership roles including committee chairs and members that acceptance of the position or award indicates that they have not had a finding or determination of sexual or other harassment or sexual misconduct against them from a current or past employer within the previous eight years. The details of this process will be worked out with LERA staff."

These two motions A and B associated with item 3 both passed, seconded by Bill Canak, and passed with one opposition.

Next meeting will take place on January 5, 2021 in Chicago, Illinois in conjunction with the LERA@ASSA Meeting.

Adjournment—The meeting was adjourned at 7:54 p.m. Eastern Time by President Dennis Dabney.

LERA General Membership Meeting Minutes
Sunday, June 14, 2020, 4:15 p.m. ET
Via Videoconference

Call to order—The meeting was called to order at 5:34 p.m. by Kris Rondeau, President. The President stated that LERA meets twice each year, once in June and once in January. Kris introduced Dennis Dabney, President Elect and Adrienne Eaton, the next President Elect. Dennis Dabney is Senior Vice President of Labor Relations at Kaiser Permanente and has 30 years of experience in HR. Adrienne Eaton is the Dean of the School of Management and Labor Relations at Rutgers University.

Committee Reports

Nominating Committee Report—Kris Rondeau reported the results of the last election. Those elected by the membership were President-Elect Adrienne Eaton from Rutgers University (2019-20 President Elect; 2020-21 President; 2021-22 Past President); academic board members were Kati Griffith, Cornell University and Jake Rosenfeld, Washington University at St. Louis; neutral/government/other board member was Heather Boushey, Washington Center for Equitable Growth; the labor board member was Erin Johansson, Jobs with Justice; the management board member was Daniel Altchek, Miles & Stockbridge P.C., and the new RVP Mid Region was Robert Chiaravalli, Strategic Labor and HR LLC. The next election will be held this summer.

Finance and Membership Report—The status of our organization in terms of membership and finance was given by Ryan Lamare. Our membership has been growing over the last several years and has stabilized at around 1,050 members. Strong attendance at the meeting (475) and consistent support from our volunteers and sponsors have resulted in a positive balance for the year. 2018 was the third consecutive year in which we had a positive balance, and 2019 is expected to be as well. Ryan thanked our sponsors, our volunteers, and our staff for these indicators of stability, and invited our members to ask their colleagues to join us.

Development and Contributions Committee Report—Jim Pruitt reports that development was strong in 2018, and in 2019 we the committee will be requesting mid-level contributions from companies and local unions. The committee also appealed to meeting attendees and members to consider making a personal contribution, depending on individual ability to do so. Jim Pruitt reiterated the importance that these personal contributions have had for the organization, and thanked everyone for their support.

Editorial Committee Report—Ariel Avgar, Editor-in-Chief, congratulated the program committee on a terrific conference. He recognized the editorial committee. The ILR Review/LERA Best Papers competition, has announced a deadline of July 15 to be considered for this year's competition. The 2019 LERA Research Volume will be published this fall, and has been edited by Susanne Bruyére. Ariel Avgar encouraged members to also ensure that their respective libraries and institutions enjoy access to LERA publications. Dionne Pohler will be editing the 2020 LERA Research Volume. The committee is currently soliciting ideas for the 2021 LERA Research Volume. The Perspectives on Work will be published this fall, and will focus on the inclusive work place. The editorial committee thanked both Mike Lillich and Bernadette Tiemann for the wonderful job they are doing with that magazine.

LERA 71st Annual Meeting Program Committee Report—Adrienne Eaton reported that the committee met yesterday to discuss the Portland meeting in 2020. The Call for Sessions is printed on the back of the LERA 70th Annual Meeting program and advertises the November 15, 2019 deadline for proposals. The theme has been selected of sustainability and the world of work, and the meetings will take place next year on Saturday, Sunday, Monday, and Tuesday. The committee is asking that those who are constructing panels to consider diversity of presenters when they are constructing panels. Regionally specific topics are also encouraged, such as marijuana and employment markets, as well as labor and environment or the forestry industry would be of interest. Bruce Kaufman has suggested that 2020 marks the 100th anniversary of the field of IR in North America, and he is invited to submit a session on this topic.

National Chapter Advisory Council Report—Bill Canak invites chapters to submit nominations for chapter awards. He reported that we have a well-functioning and successful student chapter at Rutgers University, and we are in the process of chartering a second student chapter based at Lewis and Clark University on the West Coast. We look forward to having them join us at the meeting in Portland. Bill thanked and recognized the important role of the Northeast Ohio and Central Ohio LERA Chapters in bringing this event together over the last two years. We have chartered a Virtual LERA (V LERA) chapter based at Penn State University and in their first year they have over 100 members. All their meetings and activities take place online. Bill encourages all chapter members to join LERA and all national members to join local chapters and attend their activities throughout the year. All chapter members are affiliates of the national and get access to our e-Bulletin and online publications, though they cannot vote or receive awards or other LERA member benefits. The NCAC has had some changeover in the makeup of their committee, and they have accepted a number of new members. The bylaws changes in 2016 created three new positions, three regional vice presidents specifically to provide voice for and represent LERA chapters. They each have a vote on the LERA Executive Board. Robert Chiaravalli will succeed John Budd as Mid Regional Vice President, and two nominees will run this summer to succeed Michelle Hoyman as East Regional Vice President. Those candidates will be Beverly Harrison and Tom Wassell.

2019 Awards Ceremony

LERA Media Award—Bill Canak awarded the Ken May Media Award to Moshe Z. Marvit of *The Century Foundation*; Sarah Kessler of *Quartz*; and Dave Jamieson of *The Huffington Post.*

Thomas Kochan and Stephen Sleigh Best Dissertation Award—Bruce Kauffman, BDA Chair, awarded the 2019 Best Dissertation Award to Phillippe Scrimger, Univ. of Montreal for his paper "The Distributive Effects of Trade Unionism: A Look at Income Inequality and Redistribution in Canada's Provinces". Two honorable mentions were awarded in 2019: one to Sean O'Brady, Cornell University for "Negotiating Insecurity? A Comparative Study of Collective Bargaining in Retail Food in Canada, Germany, Sweden and the United States" and one to Paula Marzionna, Cornell University, for "Is This Workplace Bullying? Conflict Management and Workplace Bullying in the Brazilian Banking Sector".

James G. Scoville Best International Paper Award—Ian Greer, BIP Chair, awarded the 2019 Best International Paper to Guglielmo Meardi for "Economic Integration and State Responses: Change in European Industrial Relations since Maastricht."

Susan C. Eaton Research Grant Award—Steve Sleigh conferred the 2019 award for the winning research proposal "Inter-firm contracting, jobs, and inequality in the U.S.: A national, regional, and industry analysis" authored by Jessica Halpern-Finnerty, UC Davis.

ILR Review/LERA Best Papers were presented by Ariel Avgar, Editor-in-Chief and Rose Batt, Editor ILRR. The papers awarded were: Tae-Youn Park, Eun-Suk Lee, and John Budd for the paper "What Do Unions Do for Mothers? Paid Maternity Leave Use and the Multifaceted Roles of Labor Unions"; Rafael Gomez and Danielle Lamb for the paper "Unions and Non-Standard Work: Union Representation and Wage Premiums across Non-Standard Work Arrangements in Canada, 1997–2014"; and Jed DeVaro, Antti Kauhanen, and Nelli Valmari, for the paper "Internal and External Hiring".

John T. Dunlop Outstanding Scholar Awards—Paul Clark presented two 2019 awards. The first award, for outstanding research addressing IR problem of national significance, was given to Maite

Tapia, Michigan State University. The second award, for exceptional contributions to international and comparative labor and employment research, was given to J. Adam Cobb, University of Texas at Austin.

The Myron C. Taylor Management Award was presented by Paul Clark. In recognition of outstanding contributions to management in the field of labor relations, this award was presented to William P. Dirksen, Ford Motor Company.

Outstanding Practitioner Award—Paul Clark presented the 2019 award, in recognition of outstanding contributions to practice in the field of labor relations, to Lisa Jordan, United Steelworkers of America and to Lu-Ann Glaser, American Water, Mid-Atlantic Division.

LERA Fellows Awards—Paul Clark presented the 2019 awards. In recognition of outstanding research and practice in the field of labor and employment relations, LERA Fellows were awarded to: John Budd, Joel Cutcher-Gershenfeld, Eileen Appelbaum, David Lewin for their respective academic contributions, and to Jim Pruitt, Owen Herrnstadt, Frances Benson, Bonnie Summers for practitioner contributions to the field of LER.

Susan C. Eaton Outstanding Scholar-Practitioner Award—Paul Clark presented the 2019 award. In recognition of outstanding research and practice emphasizing the value of bringing together the academic and practitioner communities in our field, this award was given to Mark Anner, Pennsylvania State University.

LERA Chapter Star Awards—TERRA (Bill Canak accepting), Long Island LERA (Thomas Wassel accepting), and NE Ohio LERA (Dennis Minni and Greg Szuter accepting) all received Chapter Star Awards for 2019.

New and Other Business

Kris Rondeau, President, announced the dates of the next annual meeting: LERA 72nd Annual Meeting, Portland Hilton, June 13 – 16, 2020 (Sat., Sun., Mon., Tues.), and the LERA@ASSA 2020, San Diego, CA, January 3 - 5, 2020 (with ASSA/AEA). New business and comments were requested from the floor. Canadian Industrial Relations Association announced its calls for papers with a deadline in September for their 2020 regional meeting in Toronto, Canada. The ceremonial gavel was passed to incoming president Dennis Dabney and an award was presented to the outgoing President, Kris Rondeau. The meeting was adjourned by new President Dennis Dabney at 5:45 p.m.

Audit Reports for 2018 and 2019

Labor and Employment Relations Association

Champaign, Illinois

Financial Statements

For the Years Ended December 31, 2019 and 2018

Feller & Kuester CPAs LLP
Certified Public Accountants
806 Parkland Court, Suite 1
Champaign, IL 61821
217-351-3192

TABLE OF CONTENTS

Feller & Kuester CPAs LLP

Tax - Audit - Bookkeeping
806 Parkland Court, Suite #1 - Champaign, Illinois 61821
Phone - (217) 351-3192 Fax - (217) 351-4135 Email - neal@fellerkuester.com

INDEPENDENT AUDITOR'S REPORT

To the Board of Directors of
Labor and Employment Relations Association
Champaign, Illinois

We have audited the accompanying financial statements of the Labor and Employment Relations Association (a nonprofit organization) which comprise the statements of financial position as of December 31, 2019 and 2018, and the related statements of activities, functional expenses, and cash flows for the years then ended, and the related notes to the financial statements.

Management's Responsibility for the Financial Statements

Management is responsible for the preparation and fair presentation of these financial statements in accordance with accounting principles generally accepted in the United States of America; this includes the design, implementation, and maintenance of internal control relevant to the preparation and fair presentation of financial statements that are free from material misstatement, whether due to fraud or error.

Auditor's Responsibility

Our responsibility is to express an opinion on these financial statements based on our audits. We conducted our audits in accordance with auditing standards generally accepted in the United States of America. Those standards require that we plan and perform the audit to obtain reasonable assurance about whether the financial statements are free from material misstatement.

An audit involves performing procedures to obtain audit evidence about the amounts and disclosures in the financial statements. The procedures selected depend on the auditor's judgment, including the assessment of the risks of material misstatement of the financial statements, whether due to fraud or error. In making those risk assessments, the auditor considers internal control relevant to the entity's preparation and fair presentation of the financial statements in order to design audit procedures that are appropriate in the circumstances, but not for the purposes of expressing an opinion on the effectiveness of the entity's internal control. Accordingly, we express no such opinion. An audit also includes evaluating the appropriateness of accounting policies used and the reasonableness of significant accounting estimates made by management, as well as evaluating the overall presentation of the financial statements.

We believe that the audit evidence we have obtained is sufficient and appropriate to provide a basis for our audit opinion.

Opinion

In our opinion, the financial statements referred to above present fairly, in all material respects, the financial position of the Labor and Employment Relations Association as of December 31, 2019 and 2018, and the changes in its net assets and its cash flows for the years then ended in accordance with the accounting principles generally accepted in the United States of America.

Feller & Kuester CPAs LLP

Feller & Kuester CPAs LLP
Champaign, Illinois

May 27, 2020

LABOR AND EMPLOYMENT RELATIONS ASSOCIATION
STATEMENTS OF FINANCIAL POSITION
DECEMBER 31, 2019 AND 2018

	2019	2018
ASSETS		
Current Assets		
Cash and Cash Equivalents	$ 283,130	$ 233,798
Investments Without Donor Restrictions	27,173	24,663
Accounts Receivable, Net	-	583
Prepaid Expenses	17,600	8,216
Inventory	7,264	6,374
Accrued Royalties	-	2,579
Total Current Assets	335,167	276,213
Property and Equipment, Net	4,367	1,508
Investments With Donor Restrictions	40,560	-
TOTAL ASSETS	$ 380,094	$ 277,721
LIABILITIES AND NET ASSETS		
Current Liabilities		
Accounts Payable	$ -	$ 82
Funds Held for Other Organizations	13,511	13,025
Accrued Liabilities	34,322	30,595
Dues Collected in Advance	70,789	72,754
Subscriptions Collected in Advance	12,066	12,848
Deferred Chapter and Organization Dues	5,600	5,050
Other Deferred Revenue	37,175	13,700
Total Current Liabilities	173,463	148,054
Net Assets		
Without Donor Restrictions		
Designated by the Board		
Susan C. Eaton Scholar-Practitioner Memorial Fund	23,779	26,779
Kochan-Sleigh Best Dissertation Award Fund	38,676	39,791
Jim Scoville Award Fund	3,924	4,539
John T. Dunlop Public Policy Fund	16,840	14,792
Gladys and Water Gershenfeld Publication Fund	17,300	16,988
Total Designated by the Board	100,519	102,889
Undesignated	65,552	26,778
Total Without Donor Restrictions	166,071	129,667
With Donor Restrictions	40,560	-
Total Net Assets	206,631	129,667
TOTAL LIABILITIES AND NET ASSETS	$ 380,094	$ 277,721

See Accompanying Notes.

LABOR AND EMPLOYMENT RELATIONS ASSOCIATION
STATEMENT OF ACTIVITIES
FOR THE YEAR ENDED DECEMBER 31, 2019

	Without Donor Restrictions			With Donor	
	Undesignated	**Designated**	**Total**	**Restrictions**	**Total**
Support and Revenues					
Membership Dues	$ 151,409	$ -	$ 151,409	$ -	$ 151,409
Meeting Income	187,254	-	187,254	-	187,254
Organization Dues & Sponsorships	62,492	-	62,492	-	62,492
Subscriptions	17,848	-	17,848	-	17,848
Chapter Fees	9,636	-	9,636	-	9,636
Member Contributions	28,001	-	28,001	39,200	67,201
Royalties	5,236	-	5,236	-	5,236
Ad Income	9,050	-	9,050	-	9,050
Publications	1,282	-	1,282	-	1,282
Mailing List Rental	-	-	-	-	-
Administrative Fees	70	-	70	-	70
Investment Income	-	2,510	2,510	1,360	3,870
Interest Income	865	-	865	-	865
Net Assets Released from Designation	4,880	(4,880)	-	-	-
Total Support and Revenues	478,023	(2,370)	475,653	40,560	516,213
Expenses					
Program Services					
General	198,684	-	198,684	-	198,684
Meetings	111,796	-	111,796	-	111,796
Publications	39,340	-	39,340	-	39,340
Supporting Services					
Management and General	70,353	-	70,353	-	70,353
Membership Development	19,076	-	19,076	-	19,076
Total Expenses	439,249	-	439,249	-	439,249
Change in Net Assets	38,774	(2,370)	36,404	40,560	76,964
Net Assets, Beginning of Year	26,778	102,889	129,667	-	129,667
Net Assets, End of Year	$ 65,552	$ 100,519	$ 166,071	$ 40,560	$ 206,631

See Accompanying Notes.

LABOR AND EMPLOYMENT RELATIONS ASSOCIATION
STATEMENT OF ACTIVITIES
FOR THE YEAR ENDED DECEMBER 31, 2018

	Without Donor Restrictions			With Donor	
	Undesignated	**Designated**	**Total**	**Restrictions**	**Total**
Support and Revenues					
Membership Dues	$ 151,608	$ -	$ 151,608	$ -	$ 151,608
Meeting Income	209,686	-	209,686	-	209,686
Organization Dues & Sponsorships	58,150	-	58,150	-	58,150
Subscriptions	16,770	-	16,770	-	16,770
Chapter Fees	8,338	-	8,338	-	8,338
Member Contributions	28,321	-	28,321	-	28,321
Royalties	11,718	-	11,718	-	11,718
Ad Income	7,300	-	7,300	-	7,300
Publications	34	-	34	-	34
Mailing List Rental	815	-	815	-	815
Administrative Fees	448	-	448	-	448
Investment Income	-	(46)	(46)	-	(46)
Interest Income	899	-	899	-	899
Net Assets Released from Designation	12,715	(12,715)	-	-	-
Total Support and Revenues	506,802	(12,761)	494,041	-	494,041
Expenses					
Program Services					
General	183,464	-	183,464	-	183,464
Meetings	153,143	-	153,143	-	153,143
Publications	35,694	-	35,694	-	35,694
Supporting Services					
Management and General	62,918	-	62,918	-	62,918
Membership Development	14,948	-	14,948	-	14,948
Total Expenses	450,167	-	450,167	-	450,167
Change in Net Assets	56,635	(12,761)	43,874	-	43,874
Net Assets, Beginning of Year	(29,857)	115,650	85,793	-	85,793
Net Assets, End of Year	$ 26,778	$ 102,889	$ 129,667	$ -	$ 129,667

See Accompanying Notes.

LABOR AND EMPLOYMENT RELATIONS ASSOCIATION
STATEMENT OF FUNCTIONAL EXPENSES
FOR THE YEAR ENDED DECEMBER 31, 2019

		MEETINGS			PUBLICATIONS			SUPPORTING SERVICES		
	General	Annual Conference	ASSA Meeting	Other Meetings	Perspectives	Research Volume	Other Publications	Management & General	Membership Development	Totals
Compensation	$ 129,378	$ -	$ -	$ -	$ -	$ -	$ -	$ 27,586	$ -	$ 156,964
Payroll Taxes & Fringes	51,654	-	-	-	-	-	-	9,712	-	61,366
Contract Labor	17,652	-	-	-	-	-	-	-	-	17,652
Depreciation	-	-	-	-	-	-	-	539	-	539
Insurance	-	-	-	-	-	-	-	923	-	923
Bank and Service Charges/Fulfillment	-	-	-	-	-	-	-	11,139	-	11,139
Promotion	-	6,569	120	-	-	-	-	-	1,472	8,161
Postage/Freight	-	-	68	6	2,219	2,176	-	1,079	-	5,548
Accounting/Auditing	-	-	-	-	-	-	-	5,015	-	5,015
Printing and Production	-	3,389	-	-	8,324	7,879	-	-	-	19,592
Services	-	2,940	-	-	7,679	9,515	1,176	-	-	21,310
Other Publication Costs	-	-	-	-	230	109	33	-	-	372
Meals/Receptions	-	57,501	5,901	50	-	-	-	-	-	63,452
Travel	-	5,605	2,345	580	-	-	-	-	-	8,530
Other Meeting Expenses	-	25,414	1,308	-	-	-	-	-	-	26,722
Computer Supplies/Services	-	-	-	-	-	-	-	10,265	-	10,265
Website	-	-	-	-	-	-	-	2,030	-	2,030
Office Supplies	-	-	-	-	-	-	-	2,059	-	2,059
Student & Member Awards	-	-	-	-	-	-	-	-	7,056	7,056
Fundraising Expense	-	-	-	-	-	-	-	-	5,933	5,933
Duplicating Expense	-	-	-	-	-	-	-	6	-	6
Chapter Expenses	-	-	-	-	-	-	-	-	4,594	4,594
Other Committee Expenses	-	-	-	-	-	-	-	-	21	21
Miscellaneous Office	-	-	-	-	-	-	-	-	-	-
Total	$ 198,684	$ 101,418	$ 9,742	$ 636	$ 18,452	$ 19,679	$ 1,209	$ 70,353	$ 19,076	$ 439,249

See Accompanying Notes.

LABOR AND EMPLOYMENT RELATIONS ASSOCIATION
STATEMENT OF FUNCTIONAL EXPENSES
FOR THE YEAR ENDED DECEMBER 31, 2018

		MEETINGS			PUBLICATIONS			SUPPORTING SERVICES		
	General	Annual Conference	ASSA Meeting	Other Meetings	Perspectives	Research Volume	Other Publications	Management & General	Membership Development	Totals
Compensation	$ 121,005	$ -	$ -	$ -	$ -	$ -	$ -	$ 23,354	$ -	$ 144,359
Payroll Taxes & Fringes	46,599	-	-	-	-	-	-	7,978	-	54,577
Contract Labor	15,860	-	-	-	-	-	-	-	-	15,860
Depreciation	-	-	-	-	-	-	-	539	-	539
Insurance	-	-	-	-	-	-	-	849	-	849
Bank and Service Charges/Fulfillment	-	-	-	-	-	-	-	9,073	-	9,073
Promotion	-	9,623	-	-	-	-	-	-	3,951	13,574
Postage/Freight	-	73	57	-	1,762	2,299	99	371	-	4,661
Accounting/Auditing	-	-	-	-	-	-	-	5,036	-	5,036
Printing and Production	-	2,678	95	-	6,541	7,540	459	-	-	17,313
Services	-	5,067	-	-	7,079	8,159	1,382	-	-	21,687
Other Publication Costs	-	-	-	-	285	89	-	-	-	374
Meals/Receptions	-	62,919	4,897	7,409	-	-	-	-	-	75,225
Travel	-	3,871	2,539	8,025	-	-	-	-	-	14,435
Other Meeting Expenses	-	45,510	280	100	-	-	-	-	-	45,890
Computer Supplies/Services	-	-	-	-	-	-	-	7,346	-	7,346
Website	-	-	-	-	-	-	-	5,733	-	5,733
Office Supplies	-	-	-	-	-	-	-	2,038	-	2,038
Student & Member Awards	-	-	-	-	-	-	-	-	7,032	7,032
Fundraising Expense	-	-	-	-	-	-	-	-	2,521	2,521
Duplicating Expense	-	-	-	-	-	-	-	433	-	433
Chapter Expenses	-	-	-	-	-	-	-	-	542	542
Other Committee Expenses	-	-	-	-	-	-	-	-	902	902
Miscellaneous Office	-	-	-	-	-	-	-	168	-	168
Total	$ 183,464	$ 129,741	$ 7,868	$ 15,534	$ 15,667	$ 18,087	$ 1,940	$ 62,918	$ 14,948	$ 450,167

See Accompanying Notes.

LABOR AND EMPLOYMENT RELATIONS ASSOCIATION
STATEMENTS OF CASH FLOWS
FOR THE YEARS ENDED DECEMBER 31, 2019 AND 2018

	2019	2018
CASH FLOWS FROM OPERATING ACTIVITIES		
Change in Net Assets	$ 76,964	$ 43,874
Adjustments to Reconcile Change in Net Assets to Net Cash Provided by (Used in) Operating Activities		
Depreciation Expense	539	539
Net Unrealized (Gains) Losses on Investments	(3,140)	736
(Increase) Decrease in Operating Assets		
Accounts Receivable, Net	583	(583)
Prepaid Expenses	(9,384)	(85)
Inventory	(890)	(929)
Accrued Royalties	2,579	5,185
Increase (Decrease) in Operating Liabilities		
Accounts Payable	(82)	(501)
Funds Held for Other Organizations	486	2,614
Accrued Liabilities	3,727	5,425
Dues Collected in Advance	(1,965)	4,216
Subscriptions Collected in Advance	(782)	2,757
Deferred Chapter Dues	550	(4,488)
Other Deferred Revenue	23,475	5,700
Net Cash Provided by (Used in) Operating Activities	92,660	64,460
CASH FLOWS FROM INVESTMENT ACTIVITIES		
Donated Investments	(39,200)	-
Reinvested Interest, Dividends, and Capital Gains	(730)	(690)
Purchases of Property and Equipment	(3,398)	-
Net Cash Provided by (Used in) Investment Activities	(43,328)	(690)
CASH FLOWS FROM FINANCING ACTIVITIES		
None	-	-
Net Cash Provided by (Used in) Financing Activities	-	-
Net Increase (Decrease) in Cash and Cash Equivalents	49,332	63,770
Beginning Cash and Cash Equivalents	233,798	170,028
Ending Cash and Cash Equivalents	$ 283,130	$ 233,798

See Accompanying Notes.

LABOR AND EMPLOYMENT RELATIONS ASSOCIATION
NOTES TO FINANCIAL STATEMENTS
DECEMBER 31, 2019 AND 2018

NOTE 1 – NATURE OF ACTIVITIES AND SIGNIFICANT ACCOUNTING POLICIES

Organization and Nature of Activities

The Labor and Employment Relations Association (the Organization) was founded in 1947 to encourage research in all aspects of the field of labor, employment, and the workplace. It is a nonprofit scholarly association of academic, labor, business, and neutral communities committed to the full discussion and exchange of ideas between and among its broad constituencies through meetings, publications, and its various electronic listservs and websites. The Organization's National Office is located in Champaign, Illinois and serves the association by planning conferences and meetings and publishing the various researches of its members. The main sources of support and revenues for the Organization are contributions, membership dues and subscriptions, and meeting income.

Subsequent Events

Subsequent events have been evaluated through May 27, 2020, which is the date the financial statements were available to be issued.

Accrual Basis of Reporting

The Organization has chosen to report on the accrual basis of accounting. Accordingly, revenue is recognized when earned and expenses are recognized when incurred in conformity with accounting principles generally accepted in the United States of America (GAAP). The financial statements are presented in accordance with Financial Accounting Standards Board (FASB) Accounting Standards Codification (ASC) 958 dated August 2016, and the provisions of the American Institute of Certified Public Accountants (AICPA) "Audit and Accounting Guide for Not-for-Profit Organizations" (the "Guide").

Estimates

The preparation of financial statements in conformity with GAAP requires management to make estimates and assumptions that affect certain reported amounts and disclosures. Accordingly, actual results could differ from those estimates.

Cash and Cash Equivalents

For purposes of reporting cash flows, cash and cash equivalents include all cash and highly liquid investments acquired with an original maturity date of three months or less. Since the penalties of converting certificate of deposits to cash is insignificant, all certificate of deposits have been included with cash and cash equivalents. As of December 31, 2019 and 2018, $73,346 and $78,226 of cash and cash equivalents are designated by the board and are subject to board-imposed stipulations

Investments

Investments consist of (1) intermediate term bond index funds which are carried at the fair value of the underlying assets and (2) Bayer AG (BAYRY) stock which is carried at the stated share price. Net appreciation (depreciation) in the fair value of investments, which consists of the realized gains or losses and the unrealized appreciation (depreciation) on those investments, is presented in the statement of activities in accordance with donor restrictions as investment income. Investment income is presented net of investment fees. The average cost method is primarily used to determine the basis for computing realized gains or losses.

Accounts Receivable

Accounts receivable are recorded primarily for outstanding invoices for membership dues and subscriptions. An allowance for doubtful accounts is based on an analysis of expected collection rates determined from experience. The Organization had an allowance for doubtful accounts of $0 and $0 at December 31, 2019 and 2018, respectively.

Inventory

The Organization's inventory of directories, research volumes, proceedings, and perspective magazines is carried at the lower of cost and market value. Cost is determined on the basis of first in – first out.

Property and Equipment

Property and equipment expenditures in excess of $500 are capitalized at cost. Donated property and equipment are capitalized at estimated cost or fair market value at the time of donation. Depreciation of the assets is computed using the straight-line method over their estimated useful lives. The range of estimated useful lives by type of asset is as follows:

Furniture and Equipment	5-7 years

Net Assets

Net assets of the Organization and changes therein are classified and reported as follows:

Net Assets without Donor Restrictions - Net assets that are not subject to donor-imposed restrictions and may be expended for any purpose in performing the primary objectives of the Organization. The Organization's board may designate assets without restrictions for specific operational purposes from time to time.

Net Assets with Donor Restrictions - Net assets subject to stipulations imposed by donors, and grantors. Some donor restrictions are temporary in nature; those restrictions will be met by actions of the Organization or by the passage of time. Other donor restrictions are perpetual in nature, where by the donor has stipulated the funds be maintained in perpetuity.

Contributions

Unconditional contributions are recognized when pledged and recorded as net assets without donor restrictions or net assets with donor restrictions, depending on the existence and/or nature of any donor-imposed restrictions. Conditional promises to give are recognized when the conditions on which they depend are substantially met. Gifts of cash and other assets are reported with donor restricted support if they are received with donor stipulations that limit the use of the donated assets.

When a restriction expires, that is, when a stipulated time restriction ends or a purpose restriction is accomplished, net assets with donor restrictions are reclassified to net assets without donor restrictions and reported in the statement of activities as net assets released from restrictions. Donor-restricted contributions whose restrictions are met in the same reporting period are reported as net assets without donor restriction support. Contributions restricted for the acquisition of land, buildings, and equipment are reported as net assets without donor restriction upon acquisition of the assets and the assets are placed in service.

Contributed Services and Goods

Contributed services are reported as contribution revenue and as assets or expenses only if the services create or enhance a non-financial asset (for example property and equipment) or:

- Would typically need to be purchased by the Organization if the services had not been provided by contribution.
- Require specialized skills.
- Are provided by individuals with those skills (such as accounting, financial, construction, educational, electrical, legal, medical, and other services provided by accountants, investments advisers, contractors, teachers, electricians, lawyers, doctors, and other professional and craftspeople).

For the years ended December 31, 2019 and 2018, the value of contributed services meeting the defined requirements for recognition in the financial statements as outlined above were not material and have not been recorded on the financial statements. The Organization does receive free office space from the University of Illinois. However, the value for the use of this office space has not been reported on the financial statements.

Income Tax Status

The Organization is a nonprofit association that is exempt from federal income tax under Section 501(c)(3) of the Internal Revenue Code. In addition, the Internal Revenue Service has determined that the Organization is not a private foundation as defined in Section 509(a)(1) and Section 170(b)(1)(A)(vi) of the Code.

The Organization has evaluated its exposure resulting from uncertain income tax position and determined the exposure is not material to the financial statements. In addition, the Organization is not aware of any tax position for which a significant change is reasonably possible within the next 12 months. Therefore, these financial statements do not include a liability for uncertain tax positions. Upon recognition of a liability for an uncertain tax position, the Organization would recognize interest expense and penalties in operating expenses.

The Organization files information tax returns in the U.S. federal jurisdiction and the state of Illinois. Its federal and Illinois information tax returns prior to fiscal year 2016 are closed. The Organization does not have any tax returns currently under examination by either the Internal Revenue Service (IRS) or any U.S. state jurisdiction.

Membership Dues and Advance Subscriptions Collected

Membership dues and subscriptions are assessed and recognized as revenue based on the life of the dues or subscription.

Functional Allocation of Expenses

The costs of providing the various programs and other activities have been summarized on a functional basis in the statements of activities. The statements of functional expenses present the natural classification detail of expenses by function. Accordingly, certain costs have been allocated among the programs and supporting services benefited.

Expenses which are easily and directly associated with a particular program or supporting service are charged directly to that functional area. Compensation and payroll taxes & fringes have been allocated on the basis of estimates of time and effort.

NOTE 2 – ARRANGEMENTS WITH THE UNIVERSITY OF ILLINOIS

The Organization moved its offices to the University of Illinois at the end of 1999. Under an arrangement with the University, the employees of the Organization are employed by the University. The employees' pension and benefits are part of the University's plans. The Organization then reimburses the University monthly for the cost of its employees.

The University of Illinois holds some cash for the Organization. These "claim on cash" balances were $45,519 and $1,878 as of December 31, 2019 and 2018, respectively. These balances have been included with cash and cash equivalents.

NOTE 3 – SERVICING ARRANGEMENTS TO ADMINISTER BUSINESS

On January 3, 2003, the Organization entered into a servicing arrangement to administer the business of the University Council of Industrial Relations and Human Resources Programs (UCIRHRP). These servicing duties include dues notification and collection, annual meeting arrangement and report preparation, and maintaining a data base and network communications for this separate organization. A separate financial statement is maintained for this company, but UCIRHRP's cash is maintained in the Organization's general bank account. The amount of cash belonging to UCIRHRP in the general account is included on the Organization's balance sheet as Funds Held for Other Organizations. UCIRHRP's cash balance in the Organization's custody was $7,660 at December 31, 2019 and $7,380 at December 31, 2018. The Organization is allowed to keep 20% of dues collected each year as an administrative fee.

The Organization entered into an agreement to administer the PhD Student Consortium. The Organization's duties include collecting donations and issuing stipends for eligible students. The amount of cash belonging to the PhD Student Consortium is included on the Organization's balance sheet as Funds Held for Other Organizations. PhD Student Consortium's cash balance in the Organization's custody was $5,851 at December 31, 2019 and $5,645 at December 31, 2018.

NOTE 4 – CONCENTRATIONS OF CREDIT RISK

Financial instruments that potentially subject the Organization to credit risk consist principally of checking accounts, money markets accounts, and certificates of deposits at financial institutions. However, management continuously monitors the Organization's balances at financial institutions. The balances at each bank as of December 31, 2019 and 2018 were insured by the Federal Deposit Insurance Corporation (FDIC) up to $250,000. As of December 31, 2019, $4,337 of the bank balance was not insured by FDIC. As of December 31, 2018, the total bank balance was fully insured. The "claim on cash" held with the University of Illinois as stated in Note 2 is not insured since it is not held in a financial institution. In addition, the Organization had undeposited funds of $3,309 and $5,441 at December 31, 2019 and 2018, respectively. Since these funds were not yet deposited with a financial institution it was covered by FDIC.

NOTE 5 – INVESTMENTS

FASB Codification 820, Fair Value Measurements, establishes a framework for measuring fair value. That framework provides a fair value hierarchy that prioritizes the inputs to valuation techniques used to measure fair value. The hierarch gives the highest priority to unadjusted quoted prices in active markets for identical assets or liabilities (Level 1 measurements) and the lowest priority to unobservable inputs (Level 3 measurements). The three levels of the fair value hierarchy under FASB Codification 820 are described below:

> Level 1 – Inputs to the valuation methodology are based on unadjusted quoted prices for identical assets or liabilities in active markets that the Organization has the ability to access.
>
> Level 2 – Inputs to the valuation methodology include quoted prices for similar assets or liabilities in active markets, quoted prices for identical or similar assets or liabilities in inactive markets, inputs other than quoted prices that are observable for the asset or liability, and inputs that are derived principally from or corroborated by observable market data by correlation or other means. If the asset or liability has a specified (contractual) term, the Level 2 input must be observable for substantially the full term of the asset or liability.
>
> Level 3 – Inputs to the valuation methodology are unobservable and significant to the fair value measurement.

The asset's and liability's fair value measurement level within the fair value hierarchy is based on the lowest level of any input that is significant to the fair value measurement. Valuation techniques used need to maximize the use of observable inputs and minimize the use of unobservable inputs.

The following is a description of the valuation methodologies used for assets measured at fair value. There have been no changes in the methodologies used at December 31, 2019 and 2018.

Intermediate Term Bond Funds: Valued at the NAV of shares held by the Organization at year-end based on readily determinable fair values, which are published daily and are the basis for current transactions.

Bayer AG (BAYRY) Stock: Valued at the stated share price at year-end.

The preceding methods described may produce a fair value calculation that may not be indicative of net realizable value or reflective of future fair values. Furthermore, although the Organization believes its valuation methods are appropriate and consistent with other market participants, the use of different methodologies or assumptions to determine the fair value of certain financial instruments could result in different fair value measurements at the reporting date.

The following table sets forth by level, within the fair value hierarchy, the Organization's assets at fair value as of December 31, 2019 and 2018:

Assets at Fair Value as of December 31, 2019

	Level 1	Level 2	Level 3	Total
Without Donor Restrictions				
Intermediate Term Bond Funds	$ 27,173	-	-	$ 27,173
With Donor Restrictions				
Bayer AG (BAYRY)	40,560	-	-	40,560
Total	$ 67,733	-	-	$ 67,733

Assets at Fair Value as of December 31, 2018

	Level 1	Level 2	Level 3	Total
Without Donor Restrictions				
Intermediate Term Bond Funds	$ 24,663	-	-	$ 24,663

Net investment earnings for the years ended December 31, 2019 and 2018 are summarized as follows:

	2019	2018
Dividends	$ 730	$ 690
Capital Gains Distributions	-	-
Net Unrealized Gains (Losses)	3,140	(736)
Total Investment Income	$ 3,870	$ (46)

This investment income, including unrealized gains and losses, are being reported on the statements of activity as investment income.

NOTE 6 – PROPERTY AND EQUIPMENT

As of December 31, 2019, Property and Equipment consists of:

Furniture and Equipment	$ 18,660
Less: Accumulated Depreciation	(14,293)
Property and Equipment, Net	$ 4,367

As of December 31, 2018, Property and Equipment consists of:

Furniture and Equipment	$ 15,262
Less: Accumulated Depreciation	(13,754)
Property and Equipment, Net	$ 1,508

Depreciation expense for the years ended December 31, 2019 and 2018 was $539 and $539, respectively.

NOTE 7 – NET ASSETS - BOARD DESIGNATED

Susan C. Eaton Scholar-Practitioner Memorial Fund

The Organization set up a memorial fund in honor of an author of a 1998 "Perspectives on Work" article who died on December 30, 2003. At its June 1, 2004 meeting, the Organization approved the establishment of an annual Susan C. Eaton Scholar-Practitioner award and grant to be paid to one or more qualified scholar researchers in even-numbered years or practitioners in odd-numbered years doing research in the labor and employment relations or related field. The Organization's Executive Board directed that 10% of any gifts received each year are undesignated and can be used for administrative expenses. As of December 31, 2019 and 2018, these designated funds totaled $23,779 and $26,779, respectively.

Kochan-Sleigh Best Dissertation Award Fund

On March 27, 2006, the Organization set up a designated fund with contributions from a member and matching contributions from General Electric. The fund is designated for a minimum of fifteen years and may be used to pay for the $1,000 annual best dissertation award and plaque. After the fifteen-year period, the fund may be continued or the amount remaining in the fund may be undesignated and become available to the Organization for unrestricted purposes. The Organization's Executive Board directed that 10% of any gifts received each year are undesignated and can be used for administrative expenses. As of December 31, 2019 and 2018, these designated funds totaled $38,676 and $39,791, respectively.

Jim Scoville Award Fund

On January 9, 2009, the Organization set up a designated fund with a $10,000 contribution from the University of Minnesota's Industrial Relations Center to honor a member and retiring professor. The fund was established to pay an annual award of $500 and a plaque for best paper on international and comparative employment issues. The Organization's Executive Board directed that 10% of any gifts received each year are undesignated and can be used for administrative expenses. As of December 31, 2019 and 2018, these designated funds totaled $3,924 and $4,539, respectively.

John T. Dunlop Public Policy Fund

On May 8, 2010, the Organization established a designated fund to subsidize a named John T. Dunlop Public Policy Session at future National Policy Forums and/or at the Organization annual meetings. The Organization's Executive Board directed that 10% of any gifts received each year are undesignated and can be used for administrative expenses. This fund will be so named for five or ten years, at which time the Organization's Executive Board will review the fund to determine whether to continue or modify the fund. As of December 31, 2019 and 2018, these designated funds totaled $16,840 and $14,792, respectively.

Gladys and Walter Gershenfeld Publication Fund

On May 8, 2010, the Organization established a designated fund to honor the memory of long-time members and past presidents Walter and Gladys Gershenfeld for the purpose of supporting the Organization's electronic and print publications. The fund will be so named for ten years, at which time the Organization will review the fund's purpose and uses, and whether to continue or modify the fund. The Organization's Executive Board directed that 10% of any gifts received each year are undesignated and can be used for administrative expenses. As of December 31, 2019 and 2018, these designated funds totaled $17,300 and $16,988, respectively.

NOTE 8 – NET ASSETS WITH DONOR RESTRICTIONS

Net assets with donor restrictions at December 31, 2019 are restricted for the following purposes or period:

Permanently restricted endowment gifts required to be retained by explicit donor stipulations and subject to Organization's endowment spending policy and appropriation:		
Support Operations of the Organization	$	40,560

NOTE 9 – ENDOWMENT

Permanently restricted net assets at December 31, 2019 consist of an endowment fund established in 2019 for the purpose of supporting operations of the Organization. Contributions to the endowment fund are subject to donor restrictions that stipulate the original principal of the gift is to be held and invested by the Organization indefinitely and income from the fund is to be expensed for operations. As required by GAAP, net assets associated with endowment funds are classified and reported on the existence or absence of donor-imposed restrictions.

The Organization is subject to the State Prudent Management of Institutional Funds Act (SPMIFA) and, thus, classifies amounts in its donor-restricted endowment funds as net assets with donor restrictions because those net assets are time restricted until the Board of Directors appropriates such amounts for expenditure. Most of those net assets also are subject to purpose restrictions that must be met before reclassifying those net assets to net assets without donor restrictions. The Board of Directors has interpreted SPMIFA as not requiring the maintenance of purchasing power of the original gift amount contributed to the endowment fund, unless a donor stipulates to the contrary. As a result of this interpretation, when reviewing its donor-restricted endowment funds, the Organization considers a fund to be underwater if the fair value of the fund is less than the sum of (a) the original value of initial and subsequent gift amounts donated to the fund, and (b) any accumulations to the fund that are required to be maintained in perpetuity in accordance with the direction of the applicable donor gift instrument. The Organization has interpreted SPMIFA to permit spending from underwater funds in accordance with the prudent measures required under the law.

Additionally, in accordance with SPMIFA, the Organization considers the following factors in making a determination to appropriate or accumulate donor-restricted endowment funds:

(1) The duration and preservation of the fund
(2) The purpose of the Organization and the donor-restricted endowment fund
(3) General economic conditions
(4) The possible effect of inflation or deflation
(5) The expected total return from income and the appreciation of investments
(6) Other resources of the Organization
(7) The investment policies of the Organization.

The Organization has adopted investment and spending policies for endowment assets that attempt to subject the fund to low investment risk and provide its operations with current income. Endowment assets are invested in 2,000 shares of Bayer AG (BAYRY) that are held by an investment company. The Organization seeks to build endowment assets through additional contributions. The Organization has a policy of appropriating for distribution each quarter the endowment fund's investment income from the previous quarter that is not permanently restricted, and the Organization generally expends the endowment fund's investment income for operations in the fiscal quarter following receipt. The current spending policy is not expected to allow the Organization's endowment fund to grow as a result of investment returns. This is consistent with the Organization's objectives to provide income for its operations, preserve endowment assets without subjecting them to substantial risk, and provide additional real growth through new gifts.

The composition of endowment net assets and the changes in endowment net assets as of December 31, 2019 are as follows:

	With Donor Restrictions
Endowment Net Assets, December 31, 2018	$ -
Member Contributions	39,200
Investment Income	1,360
Endowment Net Assets, December 31, 2019	$ 40,560

NOTE 10 – ACCRUED EXPENSES

As of December 31, 2019, Accrued Expenses consist of:

Accrued Payroll	$ 6,414
Accrued Benefits/Taxes	5,510
Accrued Vacation	22,398
Total	$ 34,322

As of December 31, 2018, Accrued Expenses consist of:

Accrued Payroll	$ 5,861
Accrued Benefits/Taxes	3,977
Accrued Vacation	20,757
Total	$ 30,595

NOTE 11 – LIQUIDITY AND AVAILABILITY OF FINANCIAL ASSETS

The following reflects the Organization's financial assets as of the statements of financial position date, reduced by amounts not available for general use because of contractual or donor-imposed restrictions within one year of the statement of financial position date. Amounts available include donor restricted amounts that are available for general expenditure in the following year. Amounts not available include amounts set aside for the items described in Note 7 that could be drawn upon if the Board of Directors approves that action.

	2019	2018
Fiscal Assets at Year-End		
Cash and Cash Equivalents	$ 283,130	$ 233,798
Investments	27,173	24,663
Accounts Receivable, Net	-	583
Less Contractual or Donor-Imposed Restrictions		
Board Designated Funds	(100,519)	(102,889)
Financial Assets Available to Meet Cash Needs for General Expenditure Within One Year	$ 209,784	$ 156,155

As part of the Organization's liquidity management, it has a policy to structure its financial assets to be available as its general expenditures, liabilities, and other obligations come due.

Excess cash is generally held in checking accounts, money markets accounts, and certificates of deposits until it is required for operational use.

NOTE 12 – COMMITMENTS

The Organization has signed contracts with hotels in Cleveland, Portland, and Detroit for the 2019, 2020, and 2021 Annual Meetings. If these contracts were cancelled at December 31, 2019 and 2018, the Organization would have owed $105,798 and $94,387, respectively, to the Cleveland, Portland, and Detroit hotels.

NOTE 13 – PENSION PLAN AND RETIREMENT CONTRIBUTIONS

The University of Illinois (the University) contributes to the State Universities Retirement System of Illinois (SURS), a cost-sharing multiple-employer defined benefit plan with a special funding situation whereby the State of Illinois (the State) makes substantially all actuarially determined required contributions on behalf of the participating employers. SURS was established July 21, 1941 to provide retirement annuities and other benefits for staff members and employees of state universities, certain affiliated organizations, and certain other state educational and scientific agencies and for survivors, dependents, and other beneficiaries of such employees. SURS is considered a component unit of the State of Illinois' financial reporting entity and is included in the State's financial reports as a pension trust fund. SURS is governed by Section 5/15, Chapter 40 of the *Illinois Compiled Statutes*. SURS issues a publicly available financial report that includes financial statements and required supplementary information. That report may be obtained by accessing the website at www.SURS.org.

Benefits Provided. A traditional benefit plan was established in 1941. Public Act 90-0448 enacted effective January 1, 1998, established an alternative defined benefit program known as the portable benefit package. The traditional and portable plan Tier 1 refers to members that began participation prior to January 1, 2011. Public Act 96-0889 revised the traditional and portable benefit plans for members who begin participation on or after January 1, 2011, and who do not have other eligible Illinois reciprocal system services. The revised plan is referred to as Tier 2. New employees are allowed 6 months after their date of hire to make an irrevocable election. A summary of the benefit provisions can be found in the SURS' comprehensive annual financial report (CAFR) Notes to the Financial Statements.

Contributions. The State of Illinois is primarily responsible for funding SURS on behalf of the individual employers at an actuarially determined amount. Public Act 88-0593 provides a Statutory Funding Plan consisting of two parts: (i) a ramp-up period from 1996 to 2010 and (ii) a period of contributions equal to a level percentage of the payroll of active members of the System to reach 90 percent of the total Actuarial Accrued Liability by the end of fiscal year 2045. Employer contributions from "trust, federal, and other funds" are provided under Section 15-155(b) of the Illinois Pension Code and require employers to pay contributions which are sufficient to cover the accruing normal costs on behalf of applicable employees. The employer normal cost was 12.46 percent of employee payroll during the period of July 1, 2017 to June 30, 2018. The employer normal cost was 12.29 percent of employee payroll during the period of July 1, 2018 to June 30, 2019. The employer normal cost was 13.02 percent of employee payroll during the period of July 1, 2019 to June 30, 2020. The normal cost is equal to the value of current year's pension benefit and does not include any allocation for the past unfunded liability or interest on the unfunded liability. Plan members are required to contribute 8.0 percent of their annual covered salary. The contribution requirements of plan members and employers are established and may be amended by the Illinois General Assembly. During the years ended December 31, 2019 and 2018, the Organization reimbursed the University of Illinois $21,137, $16,958, respectively, for the employer's payments to SURS for eligible employees.

In addition to providing pension benefits, the State of Illinois provides certain health, dental and life insurance benefits to annuitants. This includes annuitants of the Organization. Substantially all State employees, including the Organization's employees, may become eligible for postemployment benefits if they eventually become annuitants. Health and dental benefits include basic benefits for annuitants under the State's self-insurance plan and insurance contracts currently in force. Life insurance benefits for annuitants under age 60 are equal to their annual salary at the time of retirement; life insurance benefits for annuitants age 60 or older are limited to $5,000 per annuitant. Currently, the State does not segregate payments made to annuitants from those made to current employees for health, dental and life insurance benefits. These costs are funded by the State and are not an obligation of the Organization.

Employees of the Organization may also elect to participate in several tax deferred annuity plans and defined contribution plans. These are single employer plans under which benefits are provided to participating employees through contracts issued to each individual. Participation and the level of employee contributions are voluntary. The Organization is not required to make contributions.

LERA gratefully acknowledges the continuing support of its 2019–20 Major Sponsors and Organizational Members

Major Sustaining Sponsors

Kaiser Permanente
Ford Motor Company

Major University Sponsors

Cornell University, School of Industrial and Labor Relations
Rutgers University, School of Management and Labor Relations
University of Illinois at Urbana-Champaign, School of Labor and Employment Relations
Florida International University, College of Business
Pennsylvania State University, School of Labor and Employment Relations
Massachusetts Institute of Technology, Sloan School of Management/IWER

Organizational Members

American Federation of Teachers, AFL-CIO
American Train Dispatchers Association
Berkeley Research Group, Labor and Employment Practice
Black Hills Corporation
Boston University, Questrom School of Business
Cornell University, Scheinman Institute on Conflict Resolution
Cornell University, School of Industrial and Labor Relations
European Trade Union Institute
Federal Mediation and Conciliation Service
Florida International University, College of Business
Ford Motor Company
Kaiser Permanente
Massachusetts Institute of Technology, Sloan School of Management
Melbourne Law School, Centre for Employment and Labour Relations
Michigan State University, School of Human Resources and Labor Relations
Pennsylvania State University, School of Labor and Employment Relations
Portland General Electric
Rollins College, Department of Business and Social Entrepreneurship
Rutgers School of Management and Labor Relations
Taiwan Institute of Labor, Occupational Safety and Health
U.S. Department of Agriculture
United Auto Workers International
United Steelworkers of America
University of California, Los Angeles, Anderson School of Management
University of California, Los Angeles, Institute for Research on Labor and Employment
University of California-Berkeley, Center for Labor Research and Education
University of California-Berkeley, Institute for Research on Labor and Employment
University of Illinois at Urbana-Champaign, School of Labor and Employment Relations
University of Minnesota, Center for Human Resources and Labor Studies
University of Toronto, Centre for Industrial Relations and Human Resources
University of Wisconsin School for Workers
W. E. Upjohn Institute for Employment Research
West Virginia University, John Chambers College of Business and Economics

For more information about Organizational Membership with the LERA, please visit the LERA website: https://lera.memberclicks.net/organizational-membership

ALPHABETICAL LIST OF AUTHORS

Christine Bishop

Rob Boyle

Courtney Chicvak

Joel Cutcher-Gershenfeld

Dennis Dabney

Maria Figueroa

Martha E. Gimbel

Brandon Carlyle Grant

Heather L. Grob

Michael Hillard

Cady Landa

Yaminette Diaz Linhart

Niti Pandey

Tara M. Sinclair

You are invited to become a member of

THE LABOR AND EMPLOYMENT RELATIONS ASSOCIATION

The Labor and Employment Relations Association (LERA) was founded in 1947 by a group who felt that the growing field of industrial relations required an association in which professionally minded people from different organizations could meet. It was intended to enable all who were professionally interested in industrial relations to become better acquainted and to keep up to date with the practices and ideas at work in the field. To our knowledge there is no other organization that affords the multiparty exchange of ideas we have experienced over the years—a unique and valuable forum. After 72 years, both our academic and practitioners agree with the conviction of the founders that the encouragement, reporting, and critical discussion of research are essential if our professional field is to advance.

Our local and national membership of 5,000 includes representatives of management, unions, government; practitioners in consulting, arbitration, mediation, and law; and scholars and teachers representing many disciplines in colleges and universities in the United States and Canada, as well as abroad. Libraries and institutions interested in the publications of the Association are also invited to become subscribing members and thereby receive the same member publications. Organizational memberships in the Association are also available.

Membership dues cover publications for 12 months and entitle members to the electronic newsletter, membership directory and proceedings of the annual meeting online, and printed copies of the annual research volume, and the magazine, *Perspectives on Work* (volumes also available online). Additional online resources include Chapter Profiles and an IR/HR Degree Program Listing for the United States, Canada, and Australia.

Dues for the 2021 calendar year are listed below. International memberships require an additional $25.00 per year to cover postage costs.

Regular Member	$199.00
Contributing Member	$310.00
Emeritus Member (10 years+ membership and no longer employed)	$110.00
Student Member (full-time, limited to 4 consecutive years)	$30.00
Library Subscriber	$300.00

Tax-deductible financial contributions to the Association to support its educational activities are always welcome. You can make an online contribution at our website, leraweb.org, and choosing "Donate." If you are not already a member, we invite you to join the LERA by sending your membership application and dues payment. You can also join online at the LERA website, leraweb.org by choosing "Membership" or request an invoice. Inquiries regarding membership, meetings, and publications should be addressed to the LERA office.

LABOR AND EMPLOYMENT RELATIONS ASSOCIATION
University of Illinois at Urbana-Champaign
121 LER Building, 504 E. Armory Avenue
Champaign, IL 61820 USA
Phone: 217-333-0072 LERAoffice@illinois.edu Web: leraweb.org

Also in the
Labor and Employment Relations Association Series

Reimagining the Governance of Work and Employment
Edited by Dionne Pohler $34.95

Employment and Disability: Issues, Innovations, and Opportunities
Edited by Susanne Bruyere $34.95

No One Size Fits All: Worker Organization, Policy and Movement in a New Economic Age
Edited by Janice Fine, Linda Burnham, Kate Griffith, Minsun Ji, Victor Narro,
Steven Pitts, and Teyo Reyes $29.95

The Contradictions of Pension Fund Capitalism
Edited by Kevin Skerrett, Chris Roberts, Johanna Weststar, and Simon Archer $29.95

The Evolving Health Care Landscape: How Employees, Organizations, and Institutions Are Adapting and Innovating
Edited by Ariel C. Avgar and Timothy J. Vogus $29.95

Inequality, Uncertainty, and Opportunity: The Varied and Growing Role of Finance in Labor Relations
Edited by Christian E. Weller $29.95

The Disunited States of America: Employment Relations Systems in Conflict
Edited by David Jacobs and Peggy Kahn $29.95

Collective Bargaining Under Duress: Case Studies of Major U.S. Industries
Edited by Howard Stranger, Ann C. Frost, and Paul F. Clark $29.95

Public Jobs and Political Agendas: The Public Sector in an Era of Economic Stress
Edited by Daniel J. B. Mitchell $29.95

Employee Ownership and Shared Capitalism: New Direction in Research
Edited by Edward J. Carberry $29.95

Book orders and request for prices and order forms for the above LERA publications should be sent to:

ILR Press, Cornell Univ., P.O. Box 6525, Ithaca, NY 14851
Phone 607-277-2211, Fax 607-277-6292
Web www.cornellpress.cornell.edu
Additional volumes are available through ILR Press.

Information about ordering printed copies of the *Proceedings of the LERA Annual Meeting* can be found online at https://lera.memberclicks.net/publications. Orders for printed copies of *Perspectives on Work* and other LERA publications should be sent to the LERA Office:

LERA, 121 LER Building, 504 E. Armory Ave., MC-504, Champaign, IL 61820
Phone 217-333-0072, Fax 217-265-5130
Email LERAoffice@illinois.edu, Web leraweb.org